RFK

RFK

His Words for Our Times

Edited and Introduced by
Edwin O.Guthman &
C. Richard Allen

 For information address HarperCollins Publishers, 195 Broadway, New York, NY 10007.

HarperCollins books may be purchased for educational, business, or sales promotional use. For information please e-mail the Special Markets Department at SPsales@harpercollins.com.

FIRST HARPERLUXE EDITION

ISBN: 978-0-06-286385-0

Library of Congress Cataloging-in-Publication Data is available upon request.

18 19 20 21 22 ID/LSC 10 9 8 7 6 5 4 3 2 1

To our children

LES
EDDIE
GARY
AND
DIANE GUTHMAN

CHRISTOPHER
BRANDON
AND
MICHAEL ALLEN

and their children

~

You bear our hopes and fill our lives.

Contents

Part One: Journalist, Senate Committee Counsel, Campaign Manager (1948–1960)

Part Two:
Mr. Attorney General
(January 20, 1961–September 3, 1964)

Part Four:
The 1968 Presidential Campaign

Acknowledgments

This book exists because of Ed Guthman. He was my hero, mentor, and friend. One night in 1992 at his house in Pacific Palisades, California, he listened to my idea that there needed to be a published collection of Robert Kennedy's speeches, so that younger generations could understand him unmediated by biographers' lenses, and that the then-upcoming twenty-fifth anniversary of RFK's death offered a good chance to attract popular attention. Ed nodded and agreed, and I headed home. Barely a week later he called and asked me to come by his house on my way home from work, where he casually said, "I've called Ethel. We have her support. I talked to my agent. He needs a book proposal from us so he can start talking to publishers, and he says the time is short, so he'd like it by early next week."

More terrified than excited, in a rush I reminded Ed that I had never written a book, could not possibly pull this off while holding a full-time job, and anyway, had no idea what a book proposal was. Ed—a decorated army veteran of the Italian campaign in World War II, a Pulitzer Prize winner at age thirty, a newspaper editor beloved by a generation of journalists, #3 on Nixon's "enemies list," and a key RFK aide, whose friendship spanned the senator's life and continued on with his widow and children—just stared at me through his black-rimmed *Mad Men*–era glasses without blinking.

"Well," he said slowly, "you know what you want the book to do, don't you? You said you've been thinking about it for twenty-four years. What are you waiting for? Besides, I'll split it with you."

Thus began a crazy sprint I ran largely between 8 P.M. and 2 A.M. each night. Agent Ron Goldfarb, himself a veteran of RFK's Department of Justice, orchestrated a bidding process ultimately won by Viking Press at Penguin. Ed wrote the first two sections of the book because, as he noted, he "had been there"; I got to write the last two, covering Kennedy's Senate years and his presidential race, a four-year period of my adolescence that had made RFK a lifelong guiding influence. *RFK: Collected Speeches* came out in 1993, accompanied by an audio version with the brilliant actor and activist Martin Sheen narrating, and featuring the actual speech tapes for many of the selections in the book.

Ed continued (during what were theoretically his "retirement years") to teach journalism at the University of Southern California's Annenberg School for Communication for a total of two decades; was the first chair of the first Los Angeles City Ethics Commission; was one of three members of the special investigations board for Attorney General Reno examining the Branch Davidian confrontation with federal marshals in 1993; served as chairman of the RFK Foundation; and was the hub of a large extended family and immense cadre of friends. He was the most un-self-consciously moral man I've ever known; everyone has his or her personal North Star, and for me it was always easy: What Would Ed Do?

Ed seemed immortal; for his eightieth birthday, he asked his family to accompany him on a bicycle trip in Ireland. It took a rare and virulent disease to bring him down, and through it, I never heard him complain or seek pity. He remained tough as hell, generous beyond measure, and a patriot in the classic sense. Ed was profiled in Tom Brokaw's *The Greatest Generation*, and with his passing in 2009, we lost a true giant. History books in the end remember only a few of the real heroes of any age. In war and peace, across nearly eighty-nine years, Ed Guthman was his own unique rendition of an American hero.

Before he died, I promised Ed that I would get this book back into publication, and so this edition is entirely in his honor.

The two of us came to this project with differing perspectives, separated by thirty-six years in age and all that gap implies in experiences. Robert Kennedy had a profound influence on both of us, but from vastly disparate ranges: Ed knew him as a friend through the last two decades of Kennedy's life and served as his press secretary from 1961 to 1965, while I feel that my aspirations for America and our roles as citizens have been shaped largely by someone I never had the chance to meet.

There was no difference, however, in our aims for this volume or in the criteria we used in selecting a manageable number of Kennedy's speeches from the multitude (more than three hundred) that he delivered. We looked for historical interest, character revelation, substantive relevance, enduring theme, and pure beauty of language. We tried to ensure that we included those speeches that were considered "major addresses" at the time, as well as those so anointed by biographers and historians since then—but we could not incorporate all of those that we esteemed, and we are certain to have excluded a number prized by others.

For the texts of most of our selections, we are indebted to the John F. Kennedy Presidential Library and Museum, the repository of Robert Kennedy's papers. Generally, we used the press release version as the closest to that actually delivered. Where we had audiotape on a given speech, we amended the text to reflect any changes the tape indicated. Throughout, we have used ellipses to show where we have omitted text.

Many people have been invaluable to us in this project. It would have been impossible without the strong support of the Kennedy family. When asked to name his most important accomplishment, Robert Kennedy said, "Marrying Ethel." She was her husband's most vital supporter and frequent adviser; this volume benefits from her generous enthusiasm for our project over twenty-five years. She is the godmother of this book and so many other initiatives.

Robert and Ethel Kennedy had eleven children. In diverse ways, they each have gone into the family business, offering leadership in public and private sectors. They have made my life more interesting and considerably more fun, but more important, they have made this world a fairer, more caring, and more hopeful place. I could not have pulled off either edition of this book without Kerry, Joe (and Beth), and Chris opening doors for me and urging me on. I learned constantly from them, as well as from Kathleen, Bobby, Max, Rory, Courtney, and Doug—about their dad, about being an activist, about ceaseless competition and endless generosity, and about what grace under pressure looks like across a lifetime. And Michael, we so miss you.

Kennedy's words are the core of this book, but the brief opening essays reflecting on his legacy by notable leaders from around the world are beautiful, highly personal assertions of Kennedy's centrality to their lives and ours. I am so grateful to each of these contributors for lending their time and insights to this project, often on very tight deadlines.

The Kennedy library in Boston is part of the National Archives, a governmental agency staffed with expert and gracious men and women. In 1992, Will Johnson, Susan D'Entremont, Maura Porter, June Payne, Alan Goodrich, and Jim Cedrone provided hours of assistance. For this edition, the key figure was Laurie Austin of the audiovisual archives, whose knowledge of the RFK photograph collection is as vast as her patience. The National Archives in Maryland is the principal repository of America's history; there Matthew Heichelbech was vital in helping me find photographic treasures from their vaults. We taxpayers fund our National Archives, and our elected representatives have consistently cut support since the 1990s. It is a foolish savings; these are brilliant, devoted, and hard-working public servants, the keepers of a key part of our collective history and memory. The archives need and deserve far more resources.

Three of the people closest to Robert Kennedy in the Senate helped to provide the context for, and important details about, the various speeches from that period. Peter Edelman's shrewd intelligence is reflected in virtually every Kennedy program from 1964 on, and his handiwork is frequently seen in these speeches. His friendship and assistance are deeply appreciated. Adam Walinsky, the principal writer of many of the speeches selected from the period of 1965–1968, was exceptionally generous with his time and encyclopedic in his memory (plus his identification of the sources for some of the speech quotations saved us weeks of research). Edelman and Walinsky were Kennedy's two Senate legislative assistants. The late Frank Mankiewicz (Kennedy's press secretary from 1966 on) provided rich details of many of the critical incidents recounted herein as well as wonderful company.

For the 2018 edition, I had the chance to spend time with Lew Kaden, who went from the Senate office to the campaign, traveling with the senator in his final two months. Of Lew's many memorable stories, I learned the most about the connections between the candidate and the people of heartland America, a unifying appeal that we would benefit from today.

Our 1992 research also was assisted by then-senator Christopher Dodd, a contributor to the "legacy" section that begins the book and a ceaseless champion of so many of the principles for which Robert Kennedy fought. That original volume also was aided by Pat Walsh, Ann Connors, Herbert J. Miller Jr., Jack Tobin, and Dr. Michele Ghil.

The current edition came together thanks to a trio of research assistants: Erin Miller, Claire Lunde, and Samantha Bevins. Their work included coordinating the legacy contributions, which were made possible thanks to former Ambassadors John Emerson (my steady best friend for more than forty years) and Don Gips; Jeff Shesol (read his *Mutual Contempt*!), Patricia Duff, Billy Shore, Bill Carrick, Doane Liu, Rajiv Chandrasekaran, Ron Brownstein, Larry Irving, Kimberly Allen Hack, and Pilar Frank O'Leary. This edition benefited from the review and comments of Kimberly Allen Hack, Sarah Hoit, Greg Mahdesian, Wesley Bevins, Jackie

Emerson, Rebecca Grunfeld, and Randy Hack. Lynn Delaney, at Robert F. Kennedy Human Rights, has been my friend, sounding board, and link to the family for many years; she operates like a COO-diplomat fusion and always gets results. The original edition benefited from the efforts of Erin Callahan, Negin Mahboubi, Francine Pullman Brokaw, and Kathleen McDonough, and from suggestions from our friends Michael Mahdesian, Sheena Easton, Steve Nissen, Gary Ross, Nathan Gardels, and Bijan Pakzad. In South Africa, Dana Meistre and her sister Gail surmounted a range of obstacles to locate wonderful materials for us about Kennedy's 1966 trip there. And for this edition, our friend Aurash Mohebbi lent his photographic talents to getting me the first "head shots" I have ever liked.

This edition contains many photographs not seen by the public in fifty or more years—and in some cases, never widely published. I am especially grateful to some of the country's most talented photographers, whose generous support for this project was overwhelming, including Steve Schapiro; George Henry; Al Golub; George Ballis's widow, Maia, and agent Matt Herron (himself a noted photographer).

Ron Goldfarb represented us wonderfully in 1992. For the current volume, I am indebted to Jan Miller of Dupre Miller, who took this project on as a labor of love and inspired her colleagues, our publisher, and me. Thanks as well to her colleague Delsyia Aldana.

This book first reached the public thanks to Viking Press at Penguin, through Al Silverman, Mindy Werner, Sona Vogel, and Jim Brannigan. The edition you're holding in your hands came about because Henry Ferris, longtime executive editor at William Morrow at HarperCollins, believed in Kennedy and the relevance of this new edition. Henry was succeeded as editor by Geoff Shandler, who helped me focus this volume for new readers, particularly those born after 1968. He is an editor from the Ed Guthman old school, and his thoroughness and perceptive suggestions were invaluable. Vedika Khanna at HarperCollins was a gracious and effective liaison for me there, on many challenging topics, and Dale Rohrbaugh and Bob Castillo were expert copy editors and fact-checkers. Kate Ostrowka made it possible for us to offer the complete print material in our audiobook, which Jim Meskimen and James Lurie wonderfully voiced.

Finally, and most important I want to thank my family and Ed's, for their direct assistance in everything from brainstorming to editing to photograph selection—and for the support that made this project possible. For this edition, Michael Allen was indispensable throughout the process, particularly mastering all manners of software and hardware, and Chris Allen dedicated his considerable design skills to a series of cover layouts. Brandon Allen, Paige Allen, and Johanna Kirby were consistent sources of advice and assistance on everything from design to photograph selection to copyediting. All five were especially helpful in flagging where gaps in generational knowledge or experience called for more context—and where over the intervening years we have uncovered new facts or more appropriate descriptions. Thanks also for the consistent support of Ed's children (also my friends)—Les, Eddie, Gary, and Diane. If you find this book of value and pleasure, you have my wife, Loree, to thank, as I do every day. She encouraged me to take this project on when that meant she would have to function like a single mom for our three young kids, and she renewed the support all of these years later.

—Rick Allen, November 2017

Kennedy as Writer and Speaker

Robert Kennedy was a writer long before he was a speaker, and he never lost his insistence that words mattered. When the volume and importance of his public remarks led him to employ speech writers, the process was intense and interactive. Robert Kennedy spoke many words initially drafted by others, but he never gave a speech that was not his own.

The demands on him to speak increased in direct proportion to his emergence in the national spotlight. Prior to becoming attorney general in his brother's administration, his public speaking was limited. He gave a few talks in 1955 and 1956 about his trip with U.S. Supreme Court Justice William O. Douglas to Soviet Central Asia and Siberia; many set speeches on the work of the Senate Rackets Committee that he updated and used over and over; and frequent semi-prepared, often extemporaneous stump speeches for his brother in the 1960 presidential campaign. In those days he wrote his texts with a pen or pencil on ruled legal paper in a small, cramped hand that was difficult, and sometimes impossible, for others to decipher. The demands on his time as attorney general and the breadth of the topics he wanted (or needed) to discuss brought with them the need for help, and thereafter he had aides whose duties included—but were never limited to—drafting his speeches.

On paper Kennedy could hone a phrase or a paragraph so it would express precisely what he wanted to say. Also, he had an ear for language that was keen enough that he could recompose while speaking. He acquired a gift for eloquence that could amend a text—or go without one—and rend the heart. Kennedy's experience in the executive branch, particularly as the president's closest adviser, gave him what Adam Walinsky (one of Kennedy's Senate legislative assistants and the aide who drafted more speeches for him than anyone else) called "an inordinate faith in the fine shadings . . . just the nuance of a sentence can be everything, particularly in international affairs." When Lyndon Johnson became president, Robert Kennedy's attention to language took on an additional significance as journalists and legislators regularly searched his speeches for signs of any widening of the breach between the nation's two most visible politicians.

Despite such concerns, he still might have ended up dependent on his staff for the substantive content of his speeches had he not possessed an intellectual curiosity that grew as he matured. He went out of his way to delve firsthand into tough issues large and small, foreign and domestic—and he gained command of the subjects on which he spoke. He read widely, and he constantly reached out to people he thought could provide useful knowledge, holding seminars in his office or at his home and encouraging his aides to participate with frankness.

Frank Mankiewicz, who served as press secretary from 1966 until Kennedy's death, remembered that Kennedy encouraged him and his colleagues to gather in Kennedy's office, particularly after hours. Generally, it wasn't for small talk. "I'm convinced we need to shake up the way we look at public education," Kennedy might say. "Have you read anything good on that topic recently?" Because the staff felt compelled to try to think at least as broadly as Kennedy did, the aide being questioned most likely *would* have a book to recommend. "Can you get me a copy?" Kennedy would ask. "Do you think he [the author] would come and see me?" As likely as not, Mankiewicz said, Kennedy would have called the author the next morning before the aide got around to it.

Kennedy was comfortable surrounded by people who spoke their minds and who would push him. Unlike the hyperspecialization practiced in White House staffs of late, Kennedy never hired people only to write for him. Contrary to the situation Peggy Noonan, President Reagan's speech writer, described in her book *What I Saw at the Revolution,* Kennedy's speech writers did not feud with the "issues people"—they *were* issues people who participated in policy development and decision-making. Accordingly, they were involved with what they were writing, knew Kennedy's views in depth, and were able to write in his idiom.

A speech topic might have originated from external events, or come from Kennedy or one of his aides. Kennedy would give whoever was going to write the draft an overview of what he wanted to say. Frequently he suggested existing speeches or written documents as references or as actual drafting framework, and often he ticked off the names of people to be consulted for ideas.

In the beginning of his tenure as attorney general, Kennedy would edit heavily successive drafts prepared by his aides. As his schedule became more crowded, and his senior staff grew more comfortable that they understood his substantive desires and could write in his style, the drafting process became more streamlined, and Kennedy's modifications lessened. Still, as one of his Justice Department colleagues remembered, Kennedy went over each speech in advance, "word by word, line by line."

The process during the Senate years was similar. Kennedy was "very involved," Peter Edelman (Walinsky's colleague as Senate legislative assistant and the lead aide on domestic policy) recalled, "with reading drafts and heavily criticizing them and giving quite explicit instructions about what should be done to go to a next draft, down to the wording. One always walked away from a session like that knowing what to do. He could be very, very explicit. He could focus very well and had a good sense of language, a good ability to say 'No, that doesn't say what I mean. I mean to say something like this.'" "You have to understand," Adam Walinsky, leaning across his desk for emphasis, said nearly a quarter century after Kennedy's death, "he got to where he couldn't make it come out of his throat if he didn't believe it. He would literally lie awake at night, thinking about all of the people in all of those places around the country—the world, really—depending on him, and he took so seriously his responsibility for making things better for them."

Kennedy became a powerful speaker in the final years of his life. Partly it was because he was a Kennedy—he had picked up his fallen brother's standard and was carrying it forward—and partly it was because he had worked hard to modulate a high-pitched, somewhat raspy voice and to enunciate with penetrating clarity. But speaking in public did not come easily to him. His cousin Polly Fitzgerald, who campaigned with him in 1960, said he "forced himself to speak in those days—he was happier behind the scenes."

When he became a major national figure in the early 1960s, he was acutely aware that although he answered questions well and could move an audience extemporaneously, he was less effective when reading from text. For a few weeks in 1963 he took voice lessons from an eminent New York speech teacher, but he increasingly became bored doing the repetitive exercises she required and stopped when he got the woman's first bill and thought it was outrageously high.

Though he had a wry, needling, often self-deprecating sense of humor and uncorked it in press conferences or when speaking off-the-cuff and in spontaneous question-and-answer sessions with audiences, his formal speeches, keyed to his audiences and tightly focused on the matter that had his attention at the time, were serious, factual, straightforward, and cogent. In the vernacular of the time, "he told it like it was"; he was not flashy and did not gesture or use hyperbole very often, even as he improved his presentation. He could be eloquent, but there was an uneven quality to his speaking appearances that seemed to be linked to the chemistry of the challenge posed either by the audience or the circumstances. The greater the challenge, the better he did, as at the 1964 Democratic National Convention, at the University of Cape Town in South Africa in 1966, and at Indianapolis on the night the Reverend Martin Luther King Jr. was assassinated. Such grandeur might have surprised those who had encountered Robert Kennedy as a young man.

Introduction

Time's most common image is that of a river, with motion and direction. For most of America's two and a half centuries, its citizens and much of the world saw time's journey as flowing toward progress: improved conditions and greater happiness. Barely sixty years after the establishment of the republic, Alexis de Tocqueville observed that Americans "have all a lively faith in the perfectibility of man . . . They all consider society as a body in a state of improvement."

Across the generations, many Americans have cared little about what happened upstream; with a national history shorter than that of many countries, we keep our eyes fixed in front of us, convinced we can reach our better future even faster if we simply apply the effort.

Periodically that optimism has slipped, as if the river has spun us into an eddy, where we are constantly looping back to the same place. Or worse, as happens when the river is running furiously, time feels like a rapids' hole, into which a swimmer is swallowed and never surfaces. As I write this, America seems to many to be stuck in an eddy, and some see us being pulled under.

Few segments of the American river journey started as hopefully as the 1960s, symbolized by a young, vigorous, and glamorous president. Enormous challenges, including endemic gaps in racial and economic opportunities, remained frightening, deep, and complex—but a national spirit was strengthening in favor of change, and we had the competence and cockiness to feel certain that we could move downriver fast.

By the end of that decade, the country was deeply divided, pessimistic, and adrift. Cultural chasms over expanding rights for the marginalized, over a war that seemed endless and increasingly pointless, over even music and lifestyles, played out in politics, in entertainment, in our streets, and within many homes.

It is not the objective of this book to run that river, a personal memory now for fewer than half of us. It is instead to reach back to one of the most unique figures of that time: Robert Kennedy. He was universally known in this country and nearly as widely abroad.

He steered the election of one president, outraged that president's successor, and seemed headed to the White House himself before he was killed. He crammed more living into his forty-two years than most of us will do if given twice that time. His choice of adversaries changed dramatically, but he was consistently in battle against what he thought was wrong. Robert Kennedy expected everything of himself, drove those around him nearly as hard, and passionately insisted that each of us, and all of us, were capable of more than we ever thought possible.

Memories of Robert Kennedy have been summoned over the fifty years since his death, by biographers, painters, filmmakers, novelists, and tabloid scribblers—and by virtually every Democratic candidate for president since 1968. This book provides Robert Kennedy's most unfiltered legacy: his public statements. It is presented not as a research guide or as an exercise in nostalgia. Rather, because we believe Robert Kennedy's words summon each of us to slip the bonds of our comfort or our burdens and instead fully engage in the struggles of our age, this volume is an evocation of memory as a goad to action. Kennedy's call resonates differently, given the vagaries of readers' experiences and beliefs. For those who remember Robert Kennedy, his speeches retrace the development of what Emerson defined as a "representative man" of his time—someone who, in Kennedy biographer Arthur M. Schlesinger Jr.'s phrase, "embodies the consciousness of an epoch, who perceives things in fresh lights and new connections, who exhibits unsuspected possibilities of purpose and action to his contemporaries."

By the time he claimed his victory in California's 1968 presidential primary, Kennedy had been judged by most of those contemporaries. He spent the bulk of his adult life in public view, and he excited strong emotions in supporters and opponents for the same reason: Nearly everyone understood that he meant what he said. Reading now what he did say, after all that has occurred in each of our lives and in the story of our country since his death, gives us the chance to see the scope and depth of his views and vision in a way that may have been impossible during the chaotic 1960s. For those in the baby-boom generation, if Kennedy was a representative man of their formative years, they may measure the national and personal choices made in the last half century against the hopeful standard Kennedy's speeches proclaim—and by examining his legacy anew, they may still awaken their generation's unrealized possibilities.

But even more than to our contemporaries, this book is dedicated to those too young to remember Robert

Kennedy: to those born after 1968, and to all of our children and grandchildren. In an age where celebrity is measured by the quarter hour and heroism seldom survives the evening news, we hope young people will continue to be intrigued and inspired by a man who served only three and a half years in elective office and never became president, yet continues to be one of our most admired leaders fifty years after his murder.

In his early years, Robert Kennedy showed a talent for observation, analysis, and written expression—and a strong identification with the underdog. Clearly, he was tempered by the crises and tragedies that were to follow, but he also grew through enormous hard work and was transformed by an astonishingly broad array of experiences, which he sought out and embraced.

In addition to showing the evolution of the man, these speeches and written materials unfold an era in our recent history that continues to reverberate in today's headlines. Frequently—as when taking the "measure of a nation" at the University of Kansas in 1968; summoning the "qualities of youth" in Cape Town, South Africa, in 1966; or offering the future to those who can blend "vision, reason, and courage" later that same year at Berkeley—Kennedy's eloquence stands in the company of Churchill's and Lincoln's.

And as those leaders did, Kennedy championed

timeless principles: among them love of country, courage and the will to act, and compassion and a determination to end violence. His extemporaneous remarks the night Dr. Martin Luther King Jr. was killed, his longer meditation the next day on the tide of violence that followed that assassination, and his earlier call at Tokyo's Nihon University in 1962 for democracy and freedom throughout the world are vivid examples.

Examining Kennedy's speeches during the period between his election to the Senate and his death makes clear that Kennedy's approach to national problems did not fit neatly into the ideological categories of his time—or ours. His was a muscular liberalism, committed to an activist federal government but deeply suspicious of concentrated power and certain that fundamental change would best be achieved at the community level; insistent on responsibilities as well as rights; and convinced that the dynamism of capitalism could be the impetus for broadening national growth.

Depending on one's perspective, it is either comforting or demoralizing to recognize how many of the problems of his time remain those of our own, but it is the truth. It has become almost an axiom of politics that ideas are most valuable if they are "new," yet so many of what are regarded as the "new" ideas of contemporary politicians were posited by Kennedy more than fifty years ago. To take but one example, a single white paper (released by his campaign on May 31, 1968) included proposals for enterprise zones, portable pensions, school-site management, school testing and evaluation, health-care reform, child care, radically revised welfare that rewarded work, and a host of other ideas still being trumpeted as innovations. (The white paper can be read at rfkspeeches.com.)

We hope you will find beauty in his language, intelligence in his programs, and catalyzing hope in Robert Kennedy's vision for our country and our world. In the end, we believe he would want us to paddle hard and together on this American river journey, in the spirit of the lines of Tennyson he was fond of quoting: to be "strong in will / To strive, to seek, to find, and not to yield."

—Rick Allen, November 2017

RFK

Robert Kennedy's Legacy

The clearest insight into Robert Kennedy's character and core beliefs is to be gained by reading his speeches; hence this book. Yet Kennedy is judged great not only for what he accomplished in life but for the impact he has had on others, even after his death. We asked a number of individuals of diverse backgrounds to outline briefly their views of Kennedy's legacy. Selections follow.

Barack Obama was the forty-fourth president of the United States. The following is excerpted from remarks he gave at the Robert F. Kennedy Human Rights Award Ceremony, celebrating what would have been Senator Kennedy's eightieth birthday, on November 16, 2005.

I was only seven when Bobby Kennedy died . . . I knew him only as an icon. In that sense, it is a distance I share with most of the people who now work in this Capitol—many of whom were not even born when Bobby Kennedy died. But what's interesting is that if you go throughout the offices in the Capitol, everywhere you'll find photographs of Kennedy, or collections of his speeches, or some other memento of his life.

Why is this? Why is it that this man who was never president, who was our attorney general for only three years, who was New York's junior senator for just three and a half, still calls to us today? Still inspires our debate with his words, animates our politics with his ideas, and calls us to make gentle the life of a world that's too often coarse and unforgiving?

Obviously, much has to do with charisma and eloquence—that unique ability, rare for most but common among Kennedys, to sum up the hopes and dreams of the most diverse nation on earth with a simple phrase or sentence; to inspire even the most apathetic observers of American life.

Part of it is his youth—both the time of life and the state of mind that dared us to hope that even after John was killed; even after we lost King; there would come a younger, energetic Kennedy who could make us believe again.

But beyond these qualities, there's something more.

Within the confines of these walls and the boundaries of this city, it becomes very easy to play small-ball politics . . . [A]t some point, we stop reaching for the possible and resign ourselves to that which is most probable. This is what happens in Washington.

And yet, as this goes on, somewhere another child goes hungry in a neighborhood just blocks away from one where a family is too full to eat another bite. Somewhere another hurricane survivor still searches for a home to return to or a school for her daughter. Somewhere another twelve-year-old is gunned down by an assailant who used to be his kindergarten playmate, and another parent loses their child on the streets of Tikrit.

But somewhere, there have also always been people who believe that this isn't the way it was supposed to be—that things should be different in America. People who believe that while evil and suffering will always exist, this is a country that has been fueled by small miracles and boundless dreams—a place where we're not afraid to face down the greatest challenges in pursuit of the greater good; a place where, against all odds, we overcome.

Bobby Kennedy was one of these people.

In a nation torn by war and divided against itself, he was able to look us in the eye and tell us that no matter how many cities burned with violence, no matter how persistent the poverty or the racism, no matter how far adrift America strayed, hope would come again.

It was an idealism not based in rigid ideology. Yes, he believed that government is a force for good—but not the only force. He distrusted big bureaucracies and knew that change erupts from the will of free people in a free society; that it comes not only from new programs, but new attitudes as well. And Kennedy's was not a pie-in-the-sky-type idealism either. He believed we would always face real enemies, and that there was no quick or perfect fix to the turmoil of the 1960s.

Rather, the idealism of Robert Kennedy—the unfinished legacy that calls us still—is a fundamental belief in the continued perfection of American ideals.

It's a belief that says if this nation was truly founded on the principles of freedom and equality, it could not sit idly by while millions were shackled because of the color of their skin. That if we are to shine as a beacon of hope to the rest of the world, we must be respected not just for the might of our military, but for the reach of our ideals. That if this is a land where destiny is not determined by birth or circumstance, we have a duty to ensure that the child of a millionaire and the child of a welfare mom have the same chance in life. That if out of many, we are truly one, then we must not limit ourselves to the pursuit of selfish gain, but that which will help all Americans rise together.

We have not always lived up to these ideals and we may fail again in the future, but this legacy calls on us to try. And the reason it does—the reason we still hear the echo of not only Bobby's words, but John's and King's and Roosevelt's and Lincoln's before him—is because they stand in such stark contrast to the place in which we find ourselves today.

It's the timidity of politics that's holding us back right now—the politics of can't-do and oh-well . . . [O]ur greatness as a nation has depended on individual initiative, on a belief in the free market. But it has also depended on our sense of mutual regard for each other, the idea that everybody has a stake in the country, that we're all in it together and everybody's got a shot at opportunity.

Robert Kennedy reminded us of this. He reminds us still . . .

We don't have to accept the diminishment of the American Dream in this country now, or ever.

It's time for us to meet the whys of today with the why-nots we often quote but rarely live—to answer "why hunger" and "why homeless," "why violence" and "why despair," with "why not good jobs and living wages," "why not better health care and world-class schools," "why not a country where we make possible the potential that exists in every human being?"

If he were here today, I think it would be hard to place Robert F. Kennedy into any of the categories that so often constrain us politically. He was a fervent anti-Communist but knew diplomacy was our way out of the Cuban Missile Crisis. He sought to wage the War on Poverty but with local partnerships and community activism. He was at once both hardheaded and big-hearted.

And yet, his was not a centrism in the sense of finding a middle road or a certain point on the ideological spectrum. His was a politics that, at its heart, was deeply moral—based on the notion that in this world, there is right and there is wrong, and it's our job to organize our laws and our lives around recognizing the difference.

Bobby Kennedy spent his life making sure . . . not only to wake us from indifference and face us with the darkness we let slip into our own backyard, but to bring us the good news that we have it within our power to change all this; to write our own destiny. Because we are a people of hope. Because we are Americans.

This is the good news we still hear all these years later—the message that still points us down the road that Bobby Kennedy never finished traveling.

Bill Clinton was the forty-second president of the United States. The following is excerpted from remarks he gave on the twenty-fifth anniversary of Senator Kennedy's death, June 6, 1993.

[In June 1968,] on the eve of my college graduation, I cheered the victory of Robert Kennedy in the California primary and felt again that our country might face its problems openly, meet its challenges bravely, and go forward together. He dared us all. He dared the grieving not to retreat into despair. He dared the comfortable not to be complacent. He dared the doubting to keep going.

. . . [T]he memory of Robert Kennedy is so powerful that in a profound way we are all in two places today. We are here and now, and we are there, then.

For in Robert Kennedy we all invested our hopes and our dreams that somehow we might redeem the promise of the America we then feared we were losing, somehow we might call back the promise of President Kennedy and Martin Luther King and heal the divisions of Vietnam and the violence and pain in our own country. But I believe if Robert Kennedy were here today, he would dare us not to mourn his passing but to fulfill his promise and to be the people that he so badly wanted us all to be. He would dare us to leave yesterday and embrace tomorrow.

We remember him, almost captured in freeze-frame, standing on the hood of a car, grasping at outreached hands, black and brown and white. His promise was that the hands which reached out to him might someday actually reach out to each other. And together, those hands could make America everything that it ought to be, a nation reunited with itself and rededicated to its best ideals.

When his funeral train passed through the gritty cities of the Northeast, people from both sides of the tracks stood silent. He had earned their respect because he went to places most leaders never visit and listened to people most leaders never hear and spoke simple truth most leaders never speak.

He spoke out against neglect, but he challenged the neglected to seize their own destiny. He wanted so badly for government to act, but he did not trust bureaucracy. And he believed that government had to do things with people, not for them. He knew we had to do things together or not at all. He spoke to the sons and daughters of immigrants and the sons and daughters of sharecroppers and told them all, "As long as you stay apart from each other, you will never be what you ought to be."

He saw the world not in terms of right and left but right and wrong. And he taught us lessons that cannot be labeled except as powerful proof.

Robert Kennedy reminded us that on any day, in any place, at any time, racism is wrong, exploitation is wrong, violence is wrong, anything that denies the simple humanity and potential of any man or woman is wrong.

He touched children whose stomachs were swollen with hunger but whose eyes still sparkled with life. He marched with workers who strained their backs for poverty wages while harvesting our food. He walked down city streets with people who ached, not from work but from the lack of it. Then as now, his piercing eyes and urgent voice speak of the things we all like to think that we believe in.

When he was alive, some said he was ruthless. Some said he wasn't a real liberal, and others claimed he was a real radical. If he were here today, I think he would laugh and say they were both right. But now as we see him more clearly, we understand he was a man who was very gentle to those who were most vulnerable, very tough in the standards he kept for himself, very old-fashioned in the virtues in which he believed, and a relentless searcher for change, for growth, for the potential of heart and mind that he sought in himself and he demanded of others.

Robert Kennedy understood that the real purpose of leadership is to bring out the best in others. He believed the destiny of our nation is the sum total of all the decisions that all of us make. He often said that one person can make a difference, and each of us must try.

Some still believe we lost what is best about America when President Kennedy and Martin Luther King and Robert Kennedy were killed. But I ask you to remember, my fellow Americans, that Robert Kennedy did not lose his faith when his own brother was killed. And when Martin Luther King was killed, he gave from his heart what was perhaps his finest speech. He lifted himself from despair time after time and went back to work.

If you listen now you can hear with me his voice . . . telling everyone here, "We can do better." Today's troubles call us to do better . . .

Let us learn here once again the simple, powerful, beautiful lesson, the simple faith of Robert Kennedy: We can do better. Let us leave here no longer in two places, but once again in one only: in the here and now, with a commitment to tomorrow, the only part of our time that we can control. Let us embrace the memory of Robert Kennedy by living as he would have us live.

For the sake of his memory, of ourselves, and of all of our children and all those to come, let us believe again: We can do better.

Oscar Arias Sanchez, formerly the president of Costa Rica, was the 1987 Nobel Peace Prize laureate. He and the Arias Foundation are engaged in worldwide efforts for peace and reconciliation.

For the last century, relations within the Americas have been marked by intervals of stability and respect, alternating with periods of intervention and manipulation. It has been a century in which the United States has been for Latin America a democracy, an ally with dictators, a defender of liberty, and a facilitator of authoritarianism. These inconsistencies can be attributed to the dearth of people who have had the perspicacity to grasp fully the complexities and nuances of inter-American relations. Only a select few have been able to understand the unique and fragile relationship between ideology and human progress in the Americas. Robert Francis Kennedy was one such person.

Senator Kennedy's desire was to leave a better, more secure world for the next generation of Americans. He sought to do this by cultivating and expanding people's opportunities to improve themselves and their community. For Senator Kennedy, ideology was a means to work for the betterment of humankind.

He interpreted Latin America's social conflict not as a matter of ideology but as the consequence of injustice combined with the need for economic redistribution and political empowerment. Rather than regarding Communism as a malignant tumor involuntarily forcing populations into submission, he perceived it as the dire result of discontent and the desperate search for a better life.

Ideology did not impede Senator Kennedy's ability to empathize or think. He was a fervent believer in democracy. He knew that the power of real democracy, and the liberties provided by it, could easily overcome any Communist threat. He wanted to give democracy the tools and support to succeed—support not as a diluted manifestation of anti-Communism, but support that engenders democracy's full potential. By demanding equality as the foremost priority, Senator Kennedy's legacy represents a clear vision of how to implement democracy—a democracy that is of the people, by the people, and for the people. A democracy that enables people to grow and prosper in freedom. As a Latin American, it is difficult for me not to think how different our countries would be if Robert Francis Kennedy had lived.

[Written for the 1993 edition.]

Bono is lead singer of U2 and cofounder of the ONE Campaign and (RED). He is also the 2009 recipient of the Robert F. Kennedy Ripple of Hope Award.

For a certain kind of Irishman—and I am that certain kind—Robert F. Kennedy is our spiritual godfather.

In many homes in Ireland, the Kennedys were on the same shelf as the pope. From my father's point of view, the Kennedys were Ireland's revenge on the royal family. That's right—it took America to produce an Irish royal family.

The Kennedys had good looks and glamour, money and brains. Yes, they were Americans—no mistaking that—but they never forgot where they came from. And no Kennedy knew it more than Bobby Kennedy did. When he was a student at Harvard, he wrote a friend that "next to John F. Fitzgerald and J. P. Kennedy, I'm the toughest Irishman that lives, which makes me the toughest man that lives."

Bobby never went in for that romantic bollocks about Ireland—that nonsense notion of a sad, simple people with a gift of gab and rhyme and song and little else. Bobby's Irish were fighters. As he once said, in a speech to the Friendly Sons of St. Patrick, "Ireland's chief export has been neither potatoes nor linen, but exiles and immigrants who have fought with sword and pen for freedom around the earth."

As well as a tough guy—and capable of some brutish behavior—Bobby Kennedy was also an idealist, albeit of the hard-nosed, sometimes hardheaded kind. He never thought "pragmatism" was a dirty word. Again, you can hear it in his speeches. Especially that audacious, miraculous one he gave in South Africa in 1966:

> *We must deal with the world as it is. We must get things done. But . . . there is no basic inconsistency between ideals and realistic possibilities, no separation between the deepest desires of heart and of mind and the rational application of human effort to human problems.*

Ideas like that are the reason why Bobby Kennedy looms so large in my life. I know he was a tough guy—and capable of some brutish behavior. On my desk at my home in Dublin, I have a picture of Bobby. Beside my bed in New York, another picture. Friends imagine it's just another holy picture for a grandiose Irishman obsessed with America. Grandiose? Yes. But they are wrong about the motive.

I need to see or read Bobby Kennedy regularly because Bobby represents the reason, the duality, and the code of conduct I aspire to in the fight for justice and human rights. His words tell us *why*—or *why not*—and his actions show us *how.*

Julian Castro, the former mayor of San Antonio, was the sixteenth U.S. Secretary of Housing and Urban Development, serving under President Obama.

I once read that Bobby Kennedy always seemed to be in the process of "becoming." The young anti-Communist who grilled Jimmy Hoffa gave way to the contemplative liberal who quoted Pericles and sat with Cesar Chavez during Chavez's hunger strike. Perhaps it was his own personal growth that gave Bobby the capacity to relate to people so different from himself and to capture with his words the essence of their struggles and their aspirations.

There's one particular passage of Bobby's that I'll never forget. In a speech to New York's Citizens Union just before Christmas 1967, Bobby addressed the plight of young urban blacks smothered by racism and the poverty it perpetuated. "How overwhelming must the frustration be of this young man—this young American," Bobby told them, "who, desperately wanting to believe and half believing, finds himself locked in the slums, his education second-rate, unable to get a job, confronted by the open prejudice and subtle hostilities of a white world, and seemingly powerless to change his condition or shape his future." *Desperately wanting to believe and half believing.* So true, so insightful, and so well put.

By the late 1960s, several generations of young Americans of color had loved their country and had wanted so much for America to return their commitment with the kind of opportunity they saw others get, only to have their hopes dashed time and again. Still, as Bobby understood, they never quite gave up on America. Instead, they clung to hope.

My mother was active in the Chicano movement of the early 1970s. I recognize her and her contemporaries in Bobby's words. Most of them grew up in the barrio, frustrated by its limits. They marched, they organized, they protested and voted, because they fundamentally believed in our country and wanted to improve it.

Bobby inspired them, and five decades later, his

words and his vision continue to inspire a new generation of Americans to make our country work for everyone—and give us all reason to believe in full.

Christopher J. Dodd was a Peace Corps volunteer in the Dominican Republic in 1966 and was later elected to the U.S. House of Representatives and the U.S. Senate, becoming the longest-serving member of Congress from Connecticut. After thirty-six years of public service, he was chairman and CEO of the Motion Picture Association of America until late 2017.

In the chaotic discourse of American politics, some parcel out promises; others cash in on fear. Robert Kennedy bartered dreams. His vision of America was a model of near-childlike simplicity, a place of boundless aspirations and unfettered opportunity. But through the commotion of the 1960s he saw the America we all knew: a land of infinite possibility, torn by inequality and haunted by self-doubt.

Robert Kennedy had that rarest of qualities, a clear-eyed conviction in what he knew to be right. He spoke with a passionate intensity about the issues of the day, whether it was the tragedy of racial hatred or the wretched conditions of the American Indian, the despair of homelessness or the blight of the urban poor. He portrayed an America where justice and fairness always prevailed, and he challenged us to create that place for ourselves.

And yet he saved his best rhetoric not for the revolution he tried to shape but for the people he tried to help. In a world of increasingly fragile social alliances, he appealed to the common denominators of decency and humanity. He held out his hand to the downtrodden and the disenfranchised, throwing open the door of society to all. And he spoke of the bitter pain of racism as if the hurt were truly his, a personal betrayal of the values he held dear.

In the end, Robert Kennedy did not change the world so much as he altered forever the way we think about it. Today his legacy remains, an indelible mark on the conscience of America. "No martyr's cause has ever been stilled by an assassin's bullet," he once said. And his was not, for now and ever.

[Written for the 1993 edition.]

Marian Wright Edelman, founder and president of the Children's Defense Fund, has been an advocate for disadvantaged Americans for her entire professional life. The first African American woman admitted to the Mississippi Bar Association, she headed the NAACP office in Jackson, an epicenter of the civil rights movement. She has been awarded the Albert Schweitzer Prize for Humanitarianism, a MacArthur Foundation Prize Fellowship, the Robert F. Kennedy Lifetime Achievement Award, and the Presidential Medal of Freedom.

In April 1967, when I was working as a young civil rights lawyer for the NAACP Legal Defense Fund in Mississippi, Senators Robert Kennedy and Joseph Clark answered my plea to visit the Mississippi Delta with me and see for themselves the hungry poor, especially children, in our very rich nation—a level of hunger and malnutrition, and worse, many people did not believe could exist in America. We visited homes where the senators asked respectfully what each family had had for breakfast, lunch, or dinner the night before. Robert Kennedy opened their empty iceboxes and cupboards after asking their permission. I watched him hover—visibly moved, on a dirt floor in a dirty, dark shack out of television-camera range—over a listless baby boy with bloated belly from whom he tried in vain to get a response as he lightly touched the baby's cheeks. When we went outside again, he asked the older children clad in dirty, ragged clothes standing in front of their shack, "What did you have for breakfast?" They responded "We ain't had no breakfast yet," although it was nearly noon. And he gently touched their faces and tried to offer words of encouragement to their hopeless and helpless mothers.

From this trip and throughout the fifteen months I knew him, until his assassination, I came to associate Robert Kennedy with nonverbal, empathetic communications that conveyed far more than words. He looked straight at you and he *saw* you—and he saw suffering children. And his capacity for genuine outrage and compassion was palpable. He kept his word to try to help Mississippi's hungry children and his pushing, passion, and visibility helped set in motion a chain of events that led to major reforms. When I complained to him many weeks later about the federal delay in getting food to hungry children and mentioned I was stopping to see Dr. Martin Luther King Jr. before returning to Jackson, he said, "Tell Dr. King to bring the poor to Washington." I did. Like Dr. King, Robert Kennedy lived and died trying to address our national plagues of physical and spiritual poverty—and exemplified the kind of moral leadership our nation sorely needs right now.

Peter Edelman is currently a professor of law at Georgetown University Law Center, and from 1965 to 1968 he was a legislative assistant to Senator Robert F. Kennedy. He has served in all three branches of the federal government.

Robert Kennedy's greatest legacy in my estimation was his passionate commitment to improving the lot of the poor and the powerless in America and around the world. That commitment is rare indeed among the powerful, but even more rare was the way he went about pursuing it. This was a man who learned and communicated by seeing, hearing, and touching. He went to see places where elected officials never go, and he went to listen, not to hold forth.

And after he listened and learned, he acted. When black leaders in Brooklyn challenged him to take action specifically in their community, he placed his intense personal energy into what became the Bedford-Stuyvesant Restoration Corporation. When he went to Mississippi and saw children close to starvation in our rich land, he went the next day to demand action from the United States secretary of agriculture and pressed the issue until he died. When he met farm workers struggling for recognition for their union in California, he became their ally and their adviser, dropping everything else on more than one occasion to assist in their struggle.

If there was or could be a new politics, Robert Kennedy personified it. He defied labels, hated them, refused to be pigeonholed. He was tough and he was loving. He was deeply critical of the damage wrought by big institutions gone wrong but committed to an activist government that, pushed by empowered people, would respond to pressing problems. He understood the levers of power and would use them even as he pressed to change both the way they operated and who controlled the switches.

Passion, commitment, activism, iconoclasm, deep religious faith, a real politics of values—all of these and more are Robert Kennedy's legacy.

[Written for the 1993 edition.]

Eric Garcetti has been the mayor of Los Angeles since 2013. He is the city's youngest mayor in more than a century, first Jewish elected mayor, and second Mexican American elected mayor. Garcetti was previously an L.A. city councilman and president of the L.A. City Council.

The assassination of Robert Kennedy seems long ago, and yet, despite the passage of fifty years, Robert Kennedy and his legacy are ever present in our lives.

Democrats and others in our nation ask how we got from the pragmatic idealism of Robert Kennedy to the presidency of Donald Trump. While the answers to that question are many and complicated, most do little to inform our future. But, an understanding of Robert Kennedy and his presidential campaign does provide us with some insight.

In the eighty-two days of the Robert Kennedy presidential campaign, Senator Kennedy sought to unite Americans of every creed, color, age, and nationality. His first primary victory came in Indiana where a coalition of African Americans and working-class whites rallied to his cause, resulting in an upset win over Senator McCarthy and the sitting governor of Indiana. It's important to note that Robert Kennedy's appeal to those Indiana working-class men and women was reflected in seven counties where in 1964 the governor of Alabama, George Wallace, had his best showing.

Robert Kennedy's last primary victory and his last day were in California, our nation's most diverse state even in 1968. This time his coalition of African

Americans, young voters, and working-class whites was joined with an energized Latino electorate to win his final race.

During that California campaign, he touched on many of the themes he discussed with voters in every corner of the country. He spoke of his opposition to the Vietnam War, but he also challenged the system of student deferments that left the war to be fought by the poor, minorities, and young men from working-class families. He spoke of crime, while always including a call for justice.

Robert Kennedy was pragmatic; after all, he was the campaign manager of his brother John's presidential campaign in 1960. He was also an idealist and in the face of political advice to the contrary, he visited the country's most impoverished places, like the Mississippi Delta and eastern Kentucky. He spent time with Native Americans on their reservations and walked with Cesar Chavez to expose injustice.

Today, a campaign would avoid these areas because the "analytics" show there are not enough votes to gain in these communities.

Let's stop asking ourselves how did we lose to Donald Trump in 2016, and learn a lesson from Robert Kennedy's 1968 campaign and start talking to everyone. Let's stop writing off voters or taking them for granted and start building a coalition and crafting a message that speaks to Latinos, African Americans, Asians, millennials, and, yes, working-class white voters.

I look at the 1968 campaign of Robert Kennedy, and while it was short and happened four years before I was born, the lessons to be learned from that moment in our history are profound for our time—fifty years later. It's time to marry RFK's unique combination of pragmatism and idealism, and apply it to the challenges of today.

Gary Hart, author, lawyer, and international business adviser, is a former U.S. senator from Colorado who twice sought the Democratic Party's nomination for U.S. president.

Had he become president, there is little doubt Robert Kennedy would have restored an age of reform begun in 1961 by his brother. Following the thesis of Arthur M. Schlesinger Jr., that American history follows cycles of conservatism and reform, the Kennedy administration represented the potential for substantial reform and renewal in the wake of the quiet Eisenhower years—but before the first mad assassin's bullet. Robert Kennedy was not only a "liberal" in his commitment to social justice, economic equality, and civil rights, he was also a progressive reformer of American domestic priorities and international policies. He wanted to move the country forward, not simply more to the left.

Perhaps his most lasting and visionary contribution was to see a world beyond the East–West ideological struggle, one which acknowledged and addressed the surpassing needs of nonaligned Latin American, African, and Asian peoples. He believed that democracy was best promoted with a helping hand, not a big stick. He foresaw an age beyond ideology when nations and peoples would be drawn together by common opportunities for peace and a desire for better conditions of life for all.

Robert Kennedy was a reform prophet of a new world order. Before anyone else, he saw beyond the Cold War, the nuclear arms race, proxy wars, and East–West polarization. Like Václav Havel two decades later, he searched for a new way, a "third way" beyond ideological confrontation and toward an agenda of common humanity.

When he died, we did not need to ask for whom the bell tolled: It tolled for all of us.

[Written for the 1993 edition.]

Tom Hayden, a longtime activist and writer, was a leader of the antiwar movement in the 1960s and served as a California state assemblyman and then state senator. He passed away in 2016.

In solitude after his brother's death, Robert Kennedy underlined this passage from Emerson's "On Heroism": "When you have chosen your part, abide by it, and do not try to weakly reconcile yourself with the world . . . Adhere to your own act, and congratulate yourself if you have . . . broken the monotony of a decorous age." Like Emerson, he was a New England transcendentalist. He felt a Universal Spirit present in pain, challenge, wilderness. His deepest speech, I think, was his denunciation of the gross national product as an idolatrous measure of human achievement [University of Kansas, March 18, 1968]. Believing in a higher power meant he could touch the best in others, remain an idealist in the center of tragedy, assert his will yet accept his fate. Anyone who saw him "adhere to his own act" through the 1968 campaign, redeeming his brother, challenging his party, opposing the war, his hair flying, those hands cut and scraped, those eyes prophetic, knows they are not likely to see such a candidacy again. But in a monotonous world of polls and packaging, he still reminds us of the alternative: the politics of soul.

[Written for the 1993 edition.]

Van Jones is a CNN political commentator, the founder of Dream Corps, and a *New York Times* bestselling author.

Many things distinguished Bobby Kennedy from the men around him, but what strikes me most is the quality of his voice. In those days, a man's voice was supposed to be firm and commanding; powerful men were expected to be stoic and in control. But Bobby's voice often quivered. His brother's murder had rocked him to his emotional core—and he could not hide that fact. As a result, his speaking voice had a raw, halting quality, rare for that age; it somehow combined strength with fragility, resolve with vulnerability.

When his brother was killed, Bobby broke. And it was in his brokenness that he discovered a deeper empathy for others who were suffering. Black or white, rich or poor—his tragedy gave him the ability to see people, hear people, *feel* people differently. His own pain allowed him to take seriously the pain of others.

And he took their citizenship seriously. He decided to pull people in, while allowing *himself* to get pulled into their circumstances.

The pain of losing his brother had transformed the hard-charging, anti-Communist, mob buster and political fixer into something else. His heartbreak opened him up to the heartbreak around him, and he took on a vulnerability rarely seen in a political leader. He became a soulful man, and that soulfulness was both visible and audible.

Bobby's example gives me permission to lead with my own heart. I was born the year he died. Born into the mourning and grief that followed; born into the remnants of hope left behind. I try to carry forward his legacy in my own work, and I am grateful to Bobby for giving me permission, as a man, to speak with authentic emotion to the painful truths of our time.

J. Robert Kerrey is a former U.S. senator from Nebraska who received the nation's highest award for heroism, the congressional Medal of Honor, for his service in Vietnam.

At the urging of a friend, I read Robert Kennedy's speeches about Vietnam twenty-five years after they were written and spoken. I feel as if I have uncovered something.

Bobby Kennedy's words did not reach me when he was alive. I entered the navy in 1966 and did not follow the political arguments about the war. Driven by the need to prepare myself for battle, I was in the bubble of the trainee.

The bubble around me broke when I arrived in Vietnam and began my first combat patrols. My time in the country was brief. I was seriously wounded on March 15, 1969, and spent the final eight months of my military career at the Philadelphia Naval Hospital.

At first I believed what I was told: My wounds would heal, I would be fitted with a prosthesis and would live a relatively normal life. This explanation seemed to make sense. My body had been changed, and therefore I assumed that I simply had been changed. Later, I discovered that the person who went to war did not return home. The person I was in 1968 was killed in Vietnam.

The words of Bobby Kennedy could not reach me when he was alive. They can reach me now.

I make this mystical confession because I was lucky, very lucky. Many of the others who physically survived the war have not been as blessed . . . [In November 1992,] I attended the tenth-anniversary celebration for the Vietnam Veterans Memorial in Washington, D.C. There I saw, as I have on other occasions, many grown men who are trapped like insect specimens in the agar of their bloody memories.

Held fast and frozen, they do not know their hell: They too died in Vietnam. Their reflections are of old men blinded by their personal trauma. Their young hearts were cut away and eaten by the ravenous cannibal of modern warfare.

Robert Kennedy was a leading character in the great tragedy of Vietnam. In the beginning he was a hot-blooded prince whose enthusiasm for battle grew from the power of his warrior brother. By the end he appeared transformed. His passion had changed; his pursuits had become noble. He had been granted possession of a vision of what might be. And then in the final act of the tragedy, this imperfect savior was felled before he reached his destination.

At the moment of his death it seemed like history had stopped, as if destiny had been snatched from our grasp.

The great miracle of life is that spirit does not die. It might be easier for us if it did. For mankind's destiny and fate cannot be less and will not be more than the sum of mankind's choices. Our fate is to be tempted by imagining what might have been. And tomorrow's tragedy begins when we yield to that temptation.

"What might have been?" is a hypnotic question that dulls and deceives us. Grieving and questioning, we postpone our work and our decisions. We lower ourselves into a grave of sadness and sadly bury our expectations too.

For surely what matters today as we read and feel the words and sense the life of Robert Kennedy—particularly his experience with the Vietnam War—is what we do with our lives. Will we who are now on life's stage choose to work the narrow trail of greatness? Or will we stumble, drunk with bitter disappointment to life's curb and surrender in resignation to walking the wide and easy path of mediocrity?

The narrow trail requires effort. Bobby Kennedy—the hero who died on his way home—carried a message imploring us to try. His words have a simple beginning: Seeing is believing. If we will simply open our eyes to the world around us; if our conscience connects and accepts some responsibility for the lonely and the suffering at our doorstep; if love carries us forward to conquer our fear and doubt, then we will know the happiness Bobby Kennedy wished for America.

[Written for the 1993 edition.]

William Manchester, a professor of history at Wesleyan University, was a noted biographer whose books included *The Last Lion* (about Winston Churchill) and *American Caesar* (about Douglas MacArthur). He passed away in 2004.

Who was Bob Kennedy?

He was a passionate Manichaean who saw mankind locked in a struggle between the powers of good and the powers of evil and who, once his inner binnacle had pointed him in the right direction, would never yield—who, in fact, didn't even know *how* to yield.

He was the last attorney general to preside over a Justice Department devoted to the pursuit of justice.

He was a man of immense inner strength and stamina, with an extraordinary capacity for growth. As attorney general he was insulted and abused by black leaders, but instead of responding with outrage or standing on his dignity—as almost anyone else would have done—he sought to understand their rage and then to do something about it.

He was a Kennedy, and he never forgot it; if any member of his family was troubled, or in need of help, he was there to counsel, to support, and, if necessary, to fight for the beleaguered Kennedy. He was a fearless and prescient leader, a rich man who became the tribune of the poor, and hoisted the flag of defiance outside the citadels of privilege and prejudice.

He was a man of compassion and vision, the very best we had.

[Written for the 1993 edition.]

Peggy Noonan, 2017 Pulitzer Prize for Commentary winner and author of *What I Saw at the Revolution,* was the principal speech writer for President Reagan and occasionally wrote for President Bush.

He brought a level of passion to politics that the arena hasn't seen since he died. He believed in belief. And he had soul: he knew poems, and could recite them.

When I was thirteen years old I went with a bunch of seventh-graders to see him in a shopping center in Massapequa, Long Island. He spoke from the top of a car. Later, as it inched through the crowd, we surrounded the car and grabbed for his hand and wouldn't let go. This was back in the '64 campaign in New York when he was up against a good solid Republican incumbent who never had a chance. I think I had the impression even as a kid that Bobby wasn't naturally gifted intellectually or physically or in terms of his personality. I thought he had to work himself into being who he was, get comfortable with himself and his circumstances, and meet each day's surprises with what he thought was a certain style. Work hard enough at appearing cool under pressure and in time you'll even fool yourself, and be cool. Work hard enough at maintaining a certain humor and you just might get humorous. I think his wit was wry in part because he got the joke, understood he was fooling himself and everyone else, and also not fooling them.

My generation, the boomers, had a lot of unfocused energy, creative and otherwise. We were the biggest, healthiest, whitest-teeth Americans ever. We had a lot of yearning. And this bright, believing man, this brother of a fallen hero, tried, quite deliberately, to summon from us the faith that we could all make America better and more decent if we joined together and, with him in the lead, stormed and took that romantic promontory called . . . government.

Not too many people think that way anymore. The boomers are grown-ups; a lot of us see government as neither romantic nor even necessarily good. Many of us have a sense that the real progress made in life, for ourselves and for others, comes through personal and private struggle. We tend toward voluntary associations and endeavors. We don't like to be told. Bobby trusted government power more than we do.

But to have been young and heard his hopeful call, to have known there was a politician who didn't like the greedy and stupid but preferred in fact the modest and even strange, to have seen him exhort people and insist we can do better—to have seen that and lived through those days is to keep, always, a sense of excitement about the possibilities of American life. It was moving and unforgettable. Which is why so many of us who grabbed for his hand will never let go.

[Written for the 1993 edition.]

Juan Manuel Santos has been president of Colombia since 2010. He was awarded the Nobel Peace Prize in 2016 "for his resolute efforts to bring the country's more than 50-year-long civil war to an end." In 2017, the National Geographic Society honored President Santos for his efforts to protect Colombia's environment.

In 1968, Robert Kennedy, at my alma mater, the University of Kansas, said, "We're going to have to negotiate. We're going to have to make compromises. We're going to have to negotiate with the National Liberation Front." He was speaking about Vietnam and the need for a negotiation because it had shattered the spirit of a nation and ended the lives of too many.

Robert F. Kennedy understood that the key to peace lies in the capacity to negotiate. Those words, spoken a year before I was admitted to the University of Kansas, resonate through time. Today, when Colombia has ended more than fifty years of internal conflict, I look back and remember when I told my fellow citizens that peace was made with your enemies and not with your friends. I also told them that to reconcile as a nation we needed to reach compromises and build gilded bridges where dignity was preserved for every stakeholder.

The road from then to now has been hard; one cannot achieve what seemed impossible without the strain of time and effort. Building peace and reconciliation will be a longtime endeavor and will need the contribution of all Colombians. But the dividends of peace, for a country like Colombia, are endless. New markets will flourish, and thousands of hectares will open to the world. Nonetheless, in a world where natural resources are becoming scarce, we are the most biodiverse country per square kilometer. Protecting the environmental dividends of peace is crucial.

Robert Kennedy understood this necessity and championed during his campaign the vision of Teddy

Roosevelt and others before him. In Colombia, we believe that peace must be achieved first, and most importantly, with nature, because "the earth's resources are not a gift from our fathers but a loan from our children."

Latin America, with all its past turmoil, is a rapidly growing region of opportunities. I would like us all to recall that period around the 1800s when American professors advised their students to "look west." We are now on the verge of something similar, but professors should instead advise to look south, south of the Rio Grande. The United States should look no farther for its strongest ally; we are the front yard of its continuing development, the building blocks of a common future. The Kennedy brothers left their spirit of Camelot here; our shared values commit us to a renewed Alliance for Progress. We will continue working, as we always have, with the people of the United States for a future filled with hope.

RFK understood that compassion is one of the most valuable ingredients for peace. He said that "if we believe that we, as Americans, are bound together by a common concern for each other, then an urgent national priority is upon us." My daughter showed me this when I took the oath of office for the first time. That day, she gave me a pair of small decorative shoes and made me have them in my office, she told me to never forget when I saw them that centuries-old maxim that goes: Always put yourself in another one's shoes. To achieve peace, you must have a sense of purpose, a strong heart to overcome egos, and the ability to embrace compassion.

Peace is a never-ending quest, and ripples of hope must always be pursued. In the end, we are one people and one race, of every color, of every belief, of every preference. The name of this one people is *the world*. The name of this one race is *humanity*. If we truly understand this, if we make it part of our individual and collective awareness, then we will cut the very root of conflicts and wars.

Howard Schultz is the executive chairman of Starbucks Corporation and a 2016 recipient of the Robert F. Kennedy Ripple of Hope Award. Through the Schultz Family Foundation, he is committed to creating opportunities for people facing barriers to success.

Robert F. Kennedy elevated our shared humanity with his ability to touch souls.

As I was growing up in the public housing projects of Brooklyn, my mother had deep admiration for John F.

Kennedy. But it was Bobby who spoke to me. When he advocated for the poor, I sensed, even as a young adult, his enormous empathy for those being left behind in a country where opportunities must be for all. And when he called for an end to the unconscionable status quo of racial inequality, I heard uncompromising compassion for all human kind.

These are traits Robert Kennedy would have brought to the White House, had he arrived. As president, he would have led with his conscience. We knew this, and when he died, millions of Americans for whom he'd stood up came and stood for him. Mile after mournful mile, along a rural stretch of tracks, his funeral train passed people aligned in reverent silence, as if frozen in his time. Even then, the country had yet to realize just how much we had lost.

The moral courage Robert Kennedy embodied is not reserved for those who govern, because it is not only governments that shape the human condition. In my office, two photographs of the senator serve as reminders for all who visit to act with the courage of our convictions even as we go about business. At company meetings, I have played excerpts of the "Ripple of Hope" speech, which he delivered in South Africa in 1966 to condemn apartheid as well as the discrimination in his own country. Fifty years later, his eloquent calls to strike out against injustice remain sadly relevant. But so does his hope.

Robert F. Kennedy's legacy is for all of us. He was not a politician but a humanitarian whose values imprinted upon me what it means to lead.

Helen Suzman served for thirty-six years in the all-white South African House of Assembly, often as the lone legislator openly opposed to apartheid. The following is adapted from her book, *In No Certain Terms: A Political Memoir*, published in 1993 and reprinted here with the permission of her daughter Dr. Frances Suzman Jowell. Helen Suzman died in 2009.

I met Robert and Ethel Kennedy when they visited South Africa in 1966. It was at the height of the confrontation between the students at the liberal English-speaking universities and the government. Student leaders had been banned and forced to live a twilight existence, forbidden to attend classes or to participate in political activities or gatherings—the latter were defined as consisting of more than two people! Some students were house-arrested—confined to their homes from 8 P.M. to 6 A.M.

His visit and his passionate espousal of liberal values were immensely encouraging to those under siege. He boosted the morale of all of us who were constantly under attack and for once we were made to feel that we were on the side of angels.

What impressed me most about Bobby Kennedy was his patent sincerity. At a private party in Johannesburg he sat chatting for a long time to a disabled African. No press was present to highlight the encounter. His stamina was remarkable. I accompanied him for one day in Cape Town. He started by visiting factories at dawn to speak to the workers, held a midmorning kaffeeklatsch with a group of women, gave a lunch-hour talk to businessmen, paid an afternoon visit to a black township where he spoke to many individuals. The day ended with his addressing a mass meeting of students at the university. His inspiring closing words were very moving. He said:

"Few will have the greatness to bend history itself, but each of us can work to change a small portion of events and in the total of all those acts will be written the history of this generation. It is from numberless, diverse acts of courage and belief that human history is shaped. Each time a man stands up for an ideal or acts to improve the lot of others, or strikes out against injustice, he sends forth a tiny ripple of hope, and crossing each other from a million different centers of energy and daring, those ripples build a current which can sweep down the mightiest walls of oppression."

[Written for the 1993 edition.]

The Most Reverend **Desmond Tutu,** a social rights activist and retired Anglican archbishop of Cape Town, South Africa, was the 1984 Nobel Peace Prize laureate.

The Kennedy family, with its long-standing commitment to democracy and against apartheid in South Africa, holds a special place in the affections of South Africans striving for freedom.

Dr. Allan Boesak, a fellow church leader, and I were privileged to host Senator Edward Kennedy in our country in 1985, when he made a tremendous impact, but it was nearly twenty years earlier that South Africans first experienced the Kennedy magic.

When Robert Kennedy came to South Africa for a whirlwind visit in 1966, a liberal newspaper, the *Rand Daily Mail,* published a souvenir booklet about the tour of the man they called "the human dynamo."

The booklet is a remarkable document. There are photographs of him being mobbed by white students in Johannesburg, and him and Ethel surrounded by black crowds in Soweto. He is pictured with the widow of the Afrikaner leader who introduced apartheid. The booklet records that students at the cradle of Afrikaner nationalism, University of Stellenbosch, cheered him despite their disagreements with him. He is photographed having discussions with an Anglican bishop opposed to apartheid—standing on an airport apron, away from police bugs. A moving photo depicts Senator Kennedy with the elderly Chief Albert Lutuli, the 1960 Nobel Prize laureate, at the rural home to which the anti-apartheid leader had been banished.

The newspaper said: "Senator Robert Kennedy's visit is the best thing that has happened to South Africa for years. It is as if a window has been flung open and a gust of fresh air has swept into a room in which the atmosphere had become stale and fetid."

The love and affection Robert Kennedy inspired and the range of contacts he was able to make were extraordinary in a society experiencing deep division and intensifying oppression. It is only now, with Nelson Mandela released and negotiation on a democratic constitution having begun, that we have begun to realize the potential which he told us we had.

A cartoon in the booklet to which I have been referring depicts two candles, entitled "Hope" and "Idealism." The caption says these two qualities were

"Kindled by Kennedy." More than a quarter of a century after his visit, he is still remembered for that achievement.

[Written for the 1993 edition.]

Elie Wiesel, Nobel Laureate, was the Andrew W. Mellon Professor in the Humanities at Boston University and the author of more than fifty-seven books. He died in 2016.

I never met Robert Kennedy. I am filled with nostalgia and melancholy when I think of the man and the dreams that haunted him: dreams of social justice and of a more authentic harmony between the various ethnic groups, communities, and classes that form a people. If death had not carried him away prematurely and tragically, our century might have drawn to an end with a smile rather than a frown.

I read his speeches, the ones before and the ones after. The ones he gave before the murder of his brother are often pragmatic. The ones delivered afterward are strikingly lyrical and sometimes prophetic: They move you deeply. Is it a new man who is speaking to us? It may be a changed man; certainly a deeper one. A man who had just discovered another dimension of injustice, pain, and human folly. How could he recover from that realization?

Battered by destiny, he wanted to show his solidarity with victims everywhere. The poor, the sick, the hopeless found in him a voice to express their suffering. War and violence discovered in him an awesome adversary whom American youth listened to fervently.

As a Jew, I feel comforted when I read what he had to say about Israel and its challenges. His comments on the Holocaust reflect a raw sensitivity, a wounded heart. It is indeed a loss not to have met him.

[Written for the 1993 edition.]

Garry Wills, a Pulitzer Prize–winning author of nearly forty books, most recently, *What the Qur'an Meant,* is an emeritus professor of history at Northwestern University.

Robert Kennedy's loss was perhaps a deeper one than that of John Kennedy. His older brother was a man formed, and formed early. Robert was still in process. He was just discovering human tragedy in the mean streets and neglected byways of America. This did not soften him, but braced him to new challenge. The famed toughness, which could be reduced to mere posturing in competition with a Jimmy Hoffa, gave a jolt of hopeful defiance to his outrage over wasted lives. He saw lost children in the slums, and their plight stoked his deep creative anger. The rest of us are tough in a less worthy sense—indurated to misery, not provokable to any useful wrath.

[Written for the 1993 edition.]

PART ONE

Journalist, Senate Committee Counsel, Campaign Manager

1948–1960

The Early Years: Out of Harvard

Kennedy stepped into the public arena shortly after completing his undergraduate studies at Harvard in 1948, not as a public official or politician but as a journalist. His reporting provides the clearest insight into his thought processes before October 10, 1955, when he gave his first formal speech—a report on his trip with U.S. Supreme Court Justice William O. Douglas to Soviet Central Asia and Siberia. On March 5, 1948, he sailed on the *Queen Mary* bound for Europe and the Middle East armed with press credentials that his father, Joseph P. Kennedy, had secured for him from the *Boston Post.* He was twenty-two years old. After spending several days in London and then Cairo, Kennedy landed in Tel Aviv on March 26—amid violence, intrigue, and intense hostility. The British mandate to rule Palestine, which derived from the old League of Nations, was due to end on May 14, because the previous November the fledgling United Nations General Assembly had decided to partition the ancient land into the two states of Israel and Palestine. As the British prepared to withdraw, Arabs and Jews were fighting fiercely throughout Palestine while the United Nations Security Council dithered over how or whether to enforce partition. Kennedy reached Jerusalem in an armored car, visited a kibbutz, and interviewed people on both sides of the conflict, including soldiers of the Haganah (the Jewish paramilitary force) and right-wing Irgun Zvai Leumi guerrillas, whose violent forays against the British and Arabs had left scores dead. In early April he went to Lebanon, and then he continued through Istanbul and Athens to Rome. On May 14, as the British mandate ended, the Jews promptly proclaimed the new state of Israel with borders conforming to the United Nations partition plan. President Harry Truman formally granted diplomatic recognition, and the next day the armies of Egypt, Jordan, Syria, Iraq, and Lebanon invaded, touching off the first Arab-Israeli War. It would end fourteen months later with the invaders defeated.

Although Kennedy wrote his pieces in April, before the May declaration of Israeli independence, the *Boston Post* published them on four successive days, beginning on June 3 when a photograph of Robert Kennedy appeared on the front page alongside a story that carried a two-column headline: "British Hated by Both Sides." An italicized introduction informed readers that the article was "the first of a special series on the Palestine situation written by Robert Kennedy," and noted "Young Kennedy has been traveling through the Middle East and his first-hand observations, appearing exclusively in the *Post,* will be of considerable interest in view of the current crisis."

Kennedy began his report by observing that Arthur Balfour, England's foreign secretary during the latter half of World War I, would have made his declaration (calling for a homeland for the Jews in Palestine) clearer if he could have foreseen how much bloodshed its ambiguity was causing. "No great thought," Kennedy wrote, "was given it at the time, for Palestine was then a relatively unimportant country. There were then not the great numbers of homeless Jews and no one believed then that the permission granted for Jewish immigration would lead 30 years later to world turmoil on whether a national home should mean an autonomous national state." Then he impartially outlined the positions of the Jews and the Arabs, asserting that "it is an unfortunate fact that because there are such well-founded arguments on either side each grows more bitter toward the other. Confidence in their right increases in direct proportion to the hatred and mistrust for the other side for not acknowledging it."

A second article detailed Kennedy's encounters with security forces and ordinary Jews, and described his visit to a kibbutz. He understood that the eight hundred thousand Jews, many of them refugees from Europe, had no place else to go. "They can go into the Mediterranean Sea and get drowned," he wrote, "or they can stay and fight and perhaps get killed. They will fight and they will fight with unparalleled courage. This is their greatest and last chance. The eyes of the world are upon them and there can be no turning back."

He continued: "Their shortages in arms and numbers are more than compensated by an undying spirit that the Arabs, Iraqis, Syrians, Lebanese, Saudi Arabians, Egyptians, and those from Trans-Jordan can never have. They are a young, tough, determined nation and will fight as such."

Kennedy's third article began with a statement that there was no chance to avoid war between the Jews and the Arabs when the British left (as had in fact happened between the time the piece was written and when it was published), and opined that the sooner the British departed, the better for the Jews because the British opposed the Jewish cause and were aiding the Arabs. Then, midway through the article, he wrote that he was becoming

> *more and more conscious of the great heritage to which we as United States citizens are heirs and which we have a duty to preserve. A force motivating my writing this paper is that I believe we have failed in this duty or are in great jeopardy of doing so . . .*
>
> *Our government first decided that justice was on the Jewish side in their desire for a homeland and then it reversed its decision temporarily. Because of this action I believe we have burdened ourselves with a great responsibility in our own eyes and in the eyes of the world. We fail to live up to that responsibility if we knowingly support the British government who behind the skirts of their official position attempt to crush a cause with which they are not in accord. If the American people knew the true facts, I am certain a more honest and forthright policy would be substituted for the benefit of all.*

In the final article, Kennedy predicted that "before too long" the United States and Great Britain would be looking to a Jewish state to preserve "a toehold" for democracy in the Middle East. Some diplomats in Washington and other Western capitals were concerned that Soviet intrigue and intervention might succeed in turning the new Jewish nation into a Communist satellite. Kennedy rejected that concern out of hand:

> *That the people might accept Communism or that Communism could exist in Palestine is fantastically absurd. Communism thrives on static discontent as sin thrives on idleness. With the type of issues and people involved, that state of affairs is nonexistent. I am as certain of that as of my name . . . Communism demands allegiance to the mother country, Russia, and it is impossible to believe that people would undergo such untold suffering to replace one tyrant with another.*

The four articles not only marked his entry into the arena of public affairs but also foretold what kind of an observer and writer he would be: strong-minded, specific, and perceptive. At a time when leading diplomats in America and around the world doubted a Jewish state was needed or could survive, he wrote that the Jews would win the war and establish "a truly great modern example of the birth of a nation in dignity and self-respect." He did not anticipate that similar yearnings among Palestinians would remain unsatisfied seven decades after his visit.

Lecture on Soviet Central Asia

GEORGETOWN UNIVERSITY; WASHINGTON, D.C.

OCTOBER 10, 1955

In 1950, U.S. Supreme Court Justice Douglas was planning a trip to five Soviet Central Asian Republics—Turkmenistan, Uzbekistan, Tadzhikistan, Kirghizia, and Kazakhstan—and at Joseph Kennedy's request, he agreed to take Robert along. The elder Kennedy and Douglas, though poles apart politically, had remained close friends since 1934 when Kennedy, then chairman of the Securities and Exchange Commission, recruited Douglas, then a Yale Law School professor, to join the SEC staff. Douglas and Robert Kennedy applied for visas in January 1951. It was near the end of Joseph Stalin's malevolent regime. Their applications were acknowledged, but visas were not issued.

Kennedy served briefly in the Internal Security Division of the Department of Justice, which prosecuted espionage and subversive-activity cases. He transferred to the Criminal Division and was working on a corruption investigation in New York when he resigned in 1952 to manage his brother John's senatorial campaign that, despite the Eisenhower landslide, unseated the Republican incumbent, Henry Cabot Lodge Jr. In January 1953, Kennedy joined the staff of the Senate Government Operations Committee's Investigations Subcommittee; at the same time, the highly controversial, Red-hunting Senator Joseph R. McCarthy of Wisconsin, as ranking Republican and Kennedy family friend, was taking over as the subcommittee's chairman. Kennedy helped investigate trade between Communist China and Western Allies while American and other allied soldiers fought under the United Nations flag against Chinese troops in Korea. In May he testified before the subcommittee that two British-owned ships, flying Panamanian flags, had transported Chinese troops.

A month later, Kennedy resigned in protest and dismay over McCarthy's heavy-handed tactics (and an intense dislike of his chief counsel, Roy Cohn). The next year he returned to the subcommittee as counsel to the Democratic minority that opposed McCarthy, and he became the subcommittee's chief counsel in 1955 after the Democrats regained control of the Senate in the 1954 election.

Douglas and Kennedy, after their visa applications to visit Soviet Central Asia were not granted in 1951, applied every year without success until January 1955. Nikita Khrushchev, the Soviet Union's leader since Joseph Stalin's death almost two years earlier, was edging toward a summit conference with President Eisenhower and, perhaps as a measure of good will, visa applications were approved within a week. That summer, Kennedy—taking leave from the Senate Investigations Subcommittee and paying his own way—and Douglas spent six weeks touring the five Central Asian republics where in many areas no Americans had been allowed since the Russian Revolution. Then they made a brief detour into Siberia before meeting Kennedy's wife, Ethel, and his sisters Jean and Pat in Moscow.

He returned very distrustful that the post-Stalin era would result in real change regarding United States–Soviet relations, though he felt Communism had raised the people's standard of living. He wrote a report on his travels, which he read at Georgetown University, accompanied by slides of his photographs from the trip. It was his maiden public speech.

In relation to its size and the antiquity of its history, the West knows less of Soviet Central Asia than any part of the civilized world. Its size of over one and a half million square miles makes it an area larger than India before partition and bigger than all of Western Europe. Its population of approximately 18 million is larger than the populations of either Canada or Australia. The two cities of Samarkand and Bukhara rival Baghdad and Damascus in cultural and religious history. For a long period of time the city of Bukhara in Uzbekistan was the center of Muslim religious fanaticism, even more so than any place in the Middle East. Yet, despite these facts and this background, we know far more about the cities and countries surrounding Soviet Central Asia than we do about this area itself. Our knowledge of Persia, Afghanistan, and even Tibet is vast compared to the information we have about Kasakhstan or Kirghizia. In fact, I doubt that there are many people in the United States who have even heard of them.

This area which was traveled and established as a main trading route by Marco Polo, overrun and conquered by Alexander the Great and Genghis Khan, and controlled by Tamerlane, has had only a handful of visitors since the turn of the sixteenth century. This was due partially to its remoteness, which became more acute when the Portuguese rounded the Cape of Good Hope and the route to Asia became a sea route rather than land, and partially to the fact that visitors were just not welcome by the local inhabitants. The first visitors to the city of Bukhara, other than a handful of envoys who came down from Russia, were two Englishmen, Stoddard and Connolly. In 1841 they visited the emir of Bukhara, the local leader of that area, in order to help him train his army to fight against the Russians. The emir, being rather distrustful of foreigners, chained them to the floor of what was aptly called the Bug Pit. After two months the emir took them out and asked them to acknowledge the Muslim religion as the only true religion. By that time they had developed a dislike of the emir and refused, whereupon he beheaded them. From personal experience I know the bugs were not the only difficulty with which Stoddard and Connolly had to contend. When Justice Douglas and I visited the Bug Pit in Bukhara last summer, the temperature read 145 degrees Fahrenheit.

The Russians began their conquest of Central Asia in the 1860s. However, some of the cities such as Bukhara remained under local control until they were taken over by the Soviets during the early 1920s and the local emirs deposed. In fact, some of the members of the harem of the emir of Bukhara are still alive.

During the Russian advance into this whole area, very few outside visitors were allowed. When the Communists took over in the 1920s, the veil of isolation hardened and for a long period it has been specifically closed to Americans. We were visiting an area of the world which, if for no other reason, was made unique by its remoteness . . .

We were interested in knowing what it was that made this colonialism acceptable to the Russians when the stated Soviet policy is against colonialism of any kind.

An added reason for us wanting to visit that part of the world was that Soviet Central Asia, prior to Communist control, had been an intensely religious area. In Bukhara alone over three hundred mosques and religious schools had flourished. How had the Muslim religion fared in the face of Communist teachings that there is no God and that religion is for the backward people?

I left Washington on the twenty-seventh of July, after stocking up on pills for stomach trouble and infection and others to purify the drinking water . . . On the first of August we took a ship from Pahlevi in Iran to Baku, the Russian oil city. We were the first Americans to take this trip on the Caspian Sea . . .

We were told before we left the U.S.A. that we must expect that all our hotel rooms in Russia would have listening devices in them and that we should govern our conversations accordingly. When we arrived in Baku we found nobody was prepared for us. No arrangements had been made to procure a guide and interpreter for us on our trip. The representative of the official Soviet tourist agency, Intourist, with whom we talked, said there was nobody available and that if we wanted to leave on our tour we could go ahead and do so but that they could not assist us. They went on to say that only one of the cities in Central Asia that we intended to visit had facilities available for tourists. To no avail did we explain that we had tried for four months to bring a guide and interpreter of our own from the U.S., that we had written four or five letters to the embassy of the Soviet Union, and that they not only refused to allow us to take an interpreter but had not even acknowledged our letters; that when we telephoned to the embassy they had said there was nothing to worry about, that the Russian government through its Intourist Agency would provide a guide. With this gloomy news we went to Justice Douglas's room, but on the way we had a whispered conversation as to our strategy. There, in rather loud voices, we discussed what an obvious mistake these people in Baku were making; that although we hated to do so we would have to telephone Khrushchev to tell him of the very bad treatment we had received, and that certainly Khrushchev would be very upset with those people who were treating us in this manner. Within an hour of this private conversation in our room, the Intourist group was knocking at our door and explaining that they had just made arrangements with Moscow to have a special guide furnished to us, and that they were going to do everything possible to facilitate our forthcoming trip, and were we sure that we had everything to make us comfortable. We were convinced from then on that our conversations were always monitored . . .

I would like to talk for a few minutes on the impressions I received on this six-week trip and what I think should be our attitude toward the present Russian policy of smiles and promises. Before the United States and our allies take steps to reduce our armies or our defense commitments, as has been discussed over the past three weeks; before we liberalize for a second time the Battle Act list[1] of strategic goods; before we take any other steps of a similar nature, we had best stop and think whether these actions are warranted, based on what Russia has said and done so far . . .

I submit history has shown that it is suicidal for the United States or its allies to make major concessions to the Soviet Union without a quid pro quo. The Soviet Union's history is replete with broken treaties and agreements . . .

When we negotiate with Soviet Russia, we are dealing with a country in which the basic freedoms and rights, which we believe to be inalienable, are for its citizens, if at all, only on the terms prescribed by those in power in the Kremlin . . .

In Alma-Ata we had a conference with the Secret Police, the first that the MVD has ever granted to a foreigner. One of the first things that they told us was how much they abhorred wiretapping and the use of listening devices. Two weeks later when Adenauer [chancellor of West Germany] was in Moscow he refused to stay in the hotel suite prepared for him because his attendants had found so many listening devices in the walls. He was finally forced to sleep in a railroad car in the station . . . In connection with the prison labor battalions in the Soviet Union, we had a rather interesting experience. On the way back to Moscow from Alma-Ata we were planning to visit some of the farm areas that had just been opened by the Soviet government under the New Lands Program.

We told our guide that on the way to the new territory we wished to stop at the city of Karaganda in central Kazakhstan, where we wanted to inspect some copper and coal mines. A day or two later our guide told us that everything had been arranged but the new land area we had originally intended to visit was not as interesting as another section. We were told that this second area was about one hour away by plane, and as it made no difference to us and as we were assured we could get to Karaganda from there, we agreed. The morning that we left we were told it was about a four-hour plane trip but that it was too late for us to change our plans. Actually it took us six and a half hours, and when we arrived there we found that we had left Kazakhstan and were now in Siberia, and furthermore the local officials told us there was no possible way of getting directly back to Karaganda, which was now much farther away than it had been before we started. They advised us, however, to fly a little farther north to the large city of Novosibirsk, and from there we certainly could get transportation to Karaganda. When we got there we were told there were no planes and only very bad trains, but by that time we told them we were going to push on anyway. Ultimately, after we had worked our way back to within a reasonable distance of Karaganda, we were told that orders had come down from Moscow that we were not to be allowed to visit that city. When we arrived in Moscow, we found out that Karaganda is the site of one of the most notorious slave-labor camps within the confines of the Soviet Union.

In more subtle ways, but equally harshly, does the state dominate and control the individual and subject him to its will. For one thing, all means of livelihood and all sources of income emanate from Moscow . . .

The right of a citizen to criticize the government does not exist in the Soviet Union. The chancellor of the University of Tashkent explained this to us by saying that the government is always right, and the people of the Soviet Union realize this, so there is nothing to criticize. Similarly, the head of the University of Frunze explained that the students had no political debates because there was only one correct position on political matters, and that was the position taken by the government, so there was no purpose in discussing the wrong or other side of the question.

The right of the free press does not exist in the Soviet Union. The control of the press by the government was brought home to me on our visit to Baku, where our guide proudly made the point that there were one hundred different newspapers and magazines published in Baku for the state of Azerbaijan. A few minutes later she pointed to a building and said that the publishing of all these periodicals was done in that one building.

The right to worship God freely does not exist in the Soviet Union. In Bukhara, although there were over three hundred mosques and religious schools forty years ago, there is now but one mosque and one Muslim school. The mosque is in a bad state of disrepair and the school must serve the whole of Central Asia. In Krasnovodsk, in Ashkhabad, and in Stalinabad we found no mosques at all . . . The Muslim religion in Soviet Central Asia has been crushed by the Soviet government. People do not become atheistic or lose their interest in religion within a period of two generations. The chancellor of the University of Alma-Ata admitted to us that during the period of communizing the countries of Central Asia, there was opposition from some of the religious leaders who, therefore, had to be eliminated. A professor at the University of Frunze said they look upon people who practice religion as backward people. It is obvious they want no backward people in the Soviet Union.

Some definite material gains have been made in Soviet Central Asia. The standard of living of the people would seem higher than the standard of living of the people in the comparable areas of the Middle East. I do not doubt but that the industrialization that is taking place in the cities, the mechanization of the farms, and the compulsory education of the citizens will undoubtedly impress people visiting Soviet Central Asia from the Middle and Far East. I think it is only right, however, that they should examine a little bit the price that has been paid and is being paid to achieve some of these gains . . .

All I ask is that before we take any more drastic steps that we receive something from the Soviet Union other than a smile and a promise—a smile that could be as crooked and a promise that could be as empty as they have been in the past. We must have peaceful coexistence with Russia, but if we and our allies are weak, there will be no peace—there will be no coexistence.

Eleven days later *US News & World Report* published a fourteen-page question-and-answer interview with Kennedy in which he told about life in Soviet Central Asia. He stressed his contempt for the Soviet system and underscored his mistrust of its leaders in the following exchange:

Q.—Would you like to go back to Russia again?
A.—I had enough of it.

Q.—Would you like to live there?
A.—No. I am a firm believer in the capitalistic system.

Q.—You think with all our troubles and all our friction and all our defects, we may be better off than they are?
A.—I think anybody who doesn't think so should take a trip there.

The Senate Rackets Committee Hearings

1956–1959

In the fall of 1956, the subcommittee, with Democratic Senator John McClellan of Arkansas as chairman and Kennedy as chief counsel, was at a loose end after exposing a conflict of interest case, which caused Air Force Secretary Harold Talbott to resign, and corruption investigations among contractors and government officials in the military clothing procurement program. Clark Mollenhoff, a Washington reporter for the *Des Moines Register* and *Look* magazine, had been urging Kennedy to investigate corruption in the labor movement. Mollenhoff was on the trail of James R. Hoffa of Detroit, then the International Brotherhood of Teamsters union boss in the Midwest, and was in contact with *Seattle Times* reporters Ed Guthman and

Paul Staples, and *Portland Oregonian* reporters Wallace Turner and William Lambert, who had written about misuse of power and money involving the Teamsters International President Dave Beck, a Seattleite, and other Teamster leaders in the Pacific Northwest. Mollenhoff thought Kennedy should start his investigation there and then move into Hoffa's domain and the East Coast.

The military clothing procurement investigation had led to indications that major New York gangsters, including Albert Anastasia and John Dioguardi (a.k.a. Johnny Dio), had muscled into the labor movement on the East Coast. The subcommittee had questioned Dioguardi in a closed session in May 1956. But Kennedy was aware that two previous congressional committees had started investigations of the Teamsters in 1953 and 1954 and then backed away because, as the chairman of the 1954 hearings, Representative Wint Smith, a Kansas Republican, told reporters, "The pressure comes from way up there and I just can't talk about it any more specifically than that." Furthermore, as Kennedy would write in his first book, *The Enemy Within*, published in 1960, "I had only a vague impression of the Teamsters Union—some notion that it was big and tough, and that Dan Tobin, Beck's immediate predecessor as president, had come from my home state of

Massachusetts. I didn't know much more about it than that."

However, Kennedy conferred with Senator McClellan and they decided in November to investigate. Kennedy traveled incognito to Los Angeles, Portland, and Seattle to interview union members, businessmen, journalists, and others, while different committee investigators began making inquiries in the Midwest and on the East Coast. In a matter of weeks they had proof that Beck was stealing thousands of dollars from Teamsters Union bank accounts. In addition, they were amassing overwhelming evidence of corruption involving Hoffa and his chief lieutenants and evidence that he and other union leaders—aided and abetted by business executives, lawyers, and financiers—were in league with mobsters. In January 1957, with the Teamsters and other labor unions challenging the subcommittee's authority to investigate labor corruption, the Senate created the Select Committee on Improper Activities in Labor and Management with McClellan as chairman and Kennedy as chief counsel. Eventually known as both the Rackets Committee and as the McClellan Committee, on February 26, 1957, the committee opened the first of a series of rugged public hearings, revealing the results of one of the most sweeping congressional investigations in the nation's history. The hearings continued, ultimately calling more than fifteen hundred witnesses, with many sessions televised live and reported in great detail, well into 1959, ending after Congress enacted the Labor Reform Act. That landmark law made embezzlement of union funds a federal crime, prohibited offering or accepting bribes in labor-management relations, provided for fair, secret-ballot union elections, and required unions to file detailed financial and administrative reports with the Department of Labor.

Inland Daily Press Association

CHICAGO, ILLINOIS

OCTOBER 15, 1957

The Labor Reform act was the McClellan Committee's major accomplishment. Thirty years old when the probe started, Kennedy was propelled by the hearings onto the national scene amid controversy and high drama, and invitations to speak about the committee poured in. He wrote a basic, matter-of-fact talk for the American Society of Newspaper Editors'* annual convention in April 1957 that he memorized, polished, updated, and used well into 1959. He often gave the speech without a reading text or notes, and the words came out quickly and easily.

I appreciate very much your invitation to be here today and discuss with you some of the problems we have uncovered so far in our investigation, as well as some of the areas in which legislation is being considered.

The committee . . . has heard over 250 different witnesses in some 130 sessions. We have received over

* Since 2009, the American Society of News Editors.

sixty-five thousand letters of complaints of improper practices in labor-management relations. We feel the committee is supported by labor throughout the country, not only because . . . the AFL-CIO Ethical Practices Committee [adopted a code of ethics], but because over 80 percent of these letters are from union members.

An analysis of these complaints and of the hearings that we have held so far, as well as the events that have transpired within organized labor itself, indicate many problem areas that demand solutions. Today, I will talk of only a few.

The Jimmy Jameses of a year ago, the Dave Becks of yesterday, the Jimmy Hoffas of today, will be forgotten. However, the defects in the law that permitted their operations will return to plague us with new Becks and Hoffas unless we find the basic solutions. The success or failure of this committee lies not in Dave Beck being denied the position of president emeritus of the Teamsters, or of Jimmy Hoffa under some half dozen federal indictments being elected president.[2] It lies, rather, in our ability to arrive at solutions for problems that obviously exist, to develop the facts so that Congress will act . . .

There are at least seven areas, admittedly controversial, where there is considerable agitation by part of the public and certain members of Congress for some federal action. They are:

1. Misuse of pension and welfare funds.

2. Misuse of union funds . . .

3. Trusteeship.

4. Democratic processes/secret ballot.

5. Terrorism/violence.

6. Organization picketing.

7. Use of union money for political purposes.

There is relatively little controversy surrounding the proposition that the federal government should exercise stricter control over pension and welfare funds . . .

It was developed that Jimmy James, an official of the Laundry Workers International Union, and several of his colleagues had taken over $900,000 of the pension and welfare funds of that union. Such misappropriation is not a federal crime and the federal government has been powerless to take any action. Because of the jurisdictional problems, local governments have so far found it impossible to move. The unhappy result has been that these people are still free and have been suspended from their jobs only recently.

In the second category, the field of union funds, apart from welfare and pension plans, there are a number of obvious deficiencies in the law. During our committee hearings into the activities of Dave Beck, the president of the International Brotherhood of Teamsters, it was established by fourteen different methods of proof that Mr. Beck took, not borrowed, some $370,000 from the funds of the Western Conference of Teamsters.

Unfortunately, here also, as the law is now written, it is not a federal crime to misappropriate funds from a labor union.

We have found certain union officials who regard their unions as their own personal possession. Union funds exist, according to their theory, for the purpose of helping them and their friends. James Hoffa, for example, continued the salary of Pontiac, Michigan, Teamster officials who were friends of his after they were convicted of extortion and sent to the penitentiary. He used over $55,000 of union members' dues money to defend several union officials who were friends of his, and who were charged and later convicted of dynamiting and extorting money from employers . . .

Mr. Hoffa transferred $500,000 of Teamsters Union funds to a Florida bank to induce them to loan money on a land scheme in which Mr. Hoffa had a financial interest. Meanwhile, the promoter of the land scheme was on the payroll of the Teamsters and was receiving some $60,000 in salary and expenses.

Mr. Hoffa also loaned $50,000 to a racetrack where his partner had a string of horses; over $150,000 to a friend who then turned around and loaned him $25,000; $400,000 to an individual who was unable to get a loan from a bank because he had been arrested twenty-two times. Several thousands of dollars of Teamster funds were used to hide his brother from the police when he was wanted for armed robbery; at another time, more than $5,000 was used to find his brother's runaway wife; $200,000 of union funds were loaned to a department store which was operated by a friend but which was in financial difficulties and, at the time, on strike by a fellow AFL union—and on and on, to well over a million dollars of union funds for Mr. Hoffa and his associates.

Not only is there no law covering the misappropriation and misuse of union funds, there is also an apparent deficiency in the law in connection with the financial reports filed by unions desiring access to the services of the National Labor Relations Board . . . We learned that under present interpretation of the law, these reports do not even have to be true or accurate. As the matter now stands, therefore, some thirty-nine thousand of these reports are filed every year, and their sole value is to take up space in the government buildings.

When these New York union officials appeared before the committee, they refused to answer questions on the grounds that their answers might tend to incriminate them. Since our hearings, a number of them have been indicted for larceny and forgery.

Certainly the law should be reexamined to determine whether it is worthwhile to have these financial statements filed and, if so, whether it should not also be required that they be accurate.

In the same connection, the relationship between labor unions and the Treasury Department is being reexamined. As you know, labor unions are tax exempt. This is a privilege, not a right.

There are those who question whether the tax-exempt status of a labor organization should continue if access to its books and records is denied to the Treasury Department, as has happened in the past, or if, most important, the tax-exempt organization is being used merely as a source of illegal income for a few individuals . . .

So, within the field of union funds alone, there are many areas that need attention . . .

In the 1956 Teamster Joint Council election in New

York City, when Mr. Hoffa was attempting to gain control of the 140,000 Teamsters in that area, votes were cast in the names of people who were listed as delegates and officers of new Teamster unions established by Hoffa and Johnny Dio who did not even know they were members of a labor organization. As one example, we had a witness appear before the committee who said that somebody came into Joe Saud's bar in lower New York, where he was having a drink, and said, "Let's all go down to the amphitheater and vote." We showed him that he was an officer in the Teamsters Union at the time and had voted. He testified he never knew he was in the Teamsters Union, let alone an officer, and, unfortunately, he did not remember if he went to the amphitheater to vote as he was too drunk that day . . .

Many of the other problems which our committee has uncovered may be traced directly to lack of democratic processes within a union. Autocratic control is an invitation to corruption, a fact too well established to be ignored . . .

A fifth area which has given concern, and on which we have held hearings and which we expect to further develop, is the use of force and violence, terrorism and bombings, in labor-management affairs. From over a dozen sections of the country the committee has received information concerning major terroristic acts in the field of labor-management relations. The state and city officials in many cases seem unable to cope with the situation . . .

Another very controversial area of complaint arises from the use of union members' dues for political purposes. Why, some argue, should a union composed of 70 percent one party and 30 percent another contribute union members' dues to the one party's candidate without the consent of the membership? On the other hand, union officials maintain they must take steps to protect the interests of the union and the members, and this political action is necessary for survival.

I am sure you will agree that all of the problems which I have cited today have been worthy of discussion and consideration by our committee. However, the role of management as well as labor also deserves close scrutiny and study. The committee is aware that there is often a very narrow line between what is extortion and what is a bribe. Undoubtedly, certain employers, because of selfish financial interests, would prefer to make sweetheart contracts with union officials to avoid paying better wages to their employees . . .

A more thorough airing of this area will come when we begin the hearings . . . We expect to develop a pattern of the improper and often illegal efforts on the part of management, including some of the largest and most respected in the country, either to avoid unionization or to bring dishonest union officials to sign a sweetheart contract . . .

We need your help and support but also your ideas on what remedial action is necessary. If the press will share the committee's responsibility in this regard, we may later be able to consider the investigation a real step forward for American society—to make this country stronger, and, for ourselves and our children, even a better place in which to live.

American Trucking Associations

WASHINGTON, D.C.

OCTOBER 20, 1959

Initially, the extent of the greed, sordid dishonesty, arbitrary abuse of power, and spidery connections with organized crime that the committee found in the Teamsters and other unions and in business surprised and offended Robert Kennedy. But before long he moved from anger to concern over what the corruption was doing to America—its structure of government and its people. Early in the investigation, as Beck's thievery and the massive corruption around Hoffa became evident, Kennedy identified increasingly with the rank-and-file members of the union. He expressed concern that people with tough jobs—truck drivers, warehousemen, and the like—were being cheated and denied not only the right to have a voice in their union but often the right to a job.

Beyond that, he came to believe that if the rich, powerful mob bosses like Hoffa and their allies in labor and management were not checked they would become more powerful than the government. By 1958 he was exhorting his staff: "We have to be successful because it can't be any other way. Hoffa said every man has his price, and you have seen his corruptive influence spread to the leading citizens of the towns across the United States, the leading bankers, the leading businessmen, officials, judges, congressmen. If he can get them to do his bidding, if he can buy them, then you can see what it means to the country. Either we are going to be successful or they are going to have the country."

Seeing that law enforcement—some seventy federal and more than ten thousand local agencies operating independently—was being outmaneuvered, outwitted, and outrun, Kennedy advocated a national crime commission to centralize gathering and correlating information on major underworld figures and disseminate it to the proper authorities. Out of that evolved the policy that he instituted only hours after becoming attorney general of mobilizing all federal agencies in a coordinated effort against organized crime. That propelled the FBI from the rear to the front in the battle, and it has been the blueprint for federal anti-racketeering action that since has ebbed and flowed depending on the degree of concern at the White House and the Department of Justice.

Those elements—the rising power of the mob, the extent of corruption, and the weakness of law enforcement—were synthesized by Kennedy into concern about whether Americans had what it would take to keep those forces in check. He publicly introduced this theme on Washington's Birthday, 1958, at the University of Notre Dame and over subsequent months, reiterated it mostly to press associations and other universities, including Georgetown, Alabama, Maryland, Tufts, and Harvard. In 1959 he was in great demand to speak to chambers of commerce, bar associations, and Rotary clubs. On October 20, he spoke to the American Trucking Associations, whose members were as directly affected by the committee's revelations as any businessmen in the country.

There is a compelling need for a reevaluation of our public attitudes toward political life. The national attitude that politics is somehow a degrading occupation for which no man of intelligence or ambition should aspire is becoming too deeply ingrained in our national thinking.

There have been many jokes directed at politics. I remember the wisecrack many years ago in Boston that my grandfather John Fitzgerald would still be mayor of Boston if someone hadn't broken into the courthouse and stolen the election returns for the next ten years.

But the time is important for us to rise in defense of politics. There is no greater need than for educated men and women to point their careers toward public service as the finest and most rewarding type of life. There is a great danger, not only in politics but in many facets of our nation's social and economic life, that the ethical and moral approach has been reduced to the second rank of importance behind the twin goals of success and prosperity. If this is generally true—and there are many indications that it is—then our nation is in dire peril at a time when its foundations should be strongest.

The last two hundred years of our country's existence have been filled with instances of people risking their security and their futures for freedom and an ideal. Washington and Lincoln; the foot soldiers at Valley Forge; the men who marched up Cemetery Hill and those who stood by their guns at the summit; the Marines who fought at Belleau Wood and at Tarawa; these men did not measure their sacrifices in terms of self-reward.

The study of [the] history of our country leaves no other conclusion but that the great events of which we are proud were forged by men of toughness who put their country above self-interest—their ideals above self-profit.

In the intervening years since Valley Forge, we have

progressed materially and financially until now we are the most powerful nation in the world. But have the comforts that we have bought, the speeches that we have made to one another on national holidays extolling American characteristics of bravery and generosity, so lulled our strength of character and moral fiber that we are now completely unprepared for the problems that are facing us? Are we prepared to meet another Valley Forge or Gettysburg? There are some very disturbing signs that we are not.

Let me give you several examples.

A few years ago, a committee of the United States Senate, for which I was counsel, made a study of the degree of cooperation and collaboration of our American soldiers who were captured in Korea. The results, which have since been supported by a report of the army itself, are compelling in their portent for America. While 50 percent of the army prisoners who were captured in Korea died in prison camps or on forced marches, Turkish troops captured at the same time, and who were generally in worse physical condition because of wounds, suffered no fatalities from cold or starvation during the prison period. The explanation, disturbing as it may sound, was that the Turkish troops took care of their wounded and their sick while many of the American troops showed concern only for themselves, shunted the sick aside, and left the wounded to die in the cold.

Even more disturbing is that one out of every ten of the American soldiers informed on his fellow prisoners on at least one occasion. A third of all the army prisoners collaborated to some extent with the Chinese. The collaboration that took place was not due to any belief in Communism, because less than 1 percent of the American prisoners showed they accepted Communism to any appreciable extent. Nor, with a few notable exceptions, was their succumbing due to any torture or brainwashing.

The explanation by the army and those who have studied the situation is this: On the part of a large percentage of Americans there was a complete lack of self-discipline; there was very little real understanding of the United States—its history or its principles; and there was no firm belief in anything—the army, family, religion, or even themselves.

What happened in Korea is not something for which we can blame the army. It is a reflection on all of us. The army is the melting pot of the people of this country, and we cannot assume that we who were not there would have done better . . .

Recently, a test designed to determine physical fitness was given to U.S. schoolchildren. Sixty percent failed to measure up to the minimum standards, while the same test given to a large number of European children resulted in only 9 percent failures.

In a period of six and a half years, from July 1950 through 1957, 5.2 million young men were called up by the draft and were examined by the armed forces, and 2.1 million of these were found to be unfit for military duty because of physical and mental deficiencies . . .

What has happened to our vital economic system—both on the side of organized labor and in business—is also of a shocking nature. The continuing survival of America depends on a vibrant economy, yet, today, we witness clear symptoms that all is not well in this vital facet of our national life.

Organized labor has undergone its own history of ultimate sacrifice on behalf of the working man. Men and women have fought and died so that their fellow employees could throw off the yoke of economic slavery. Even now the ideals of the labor movement are carried on with great credit by the vast majority of officials who follow the leadership of George Meany. However, that there are those in high positions who have betrayed their trusts, there can be no doubt.

The theft of millions of dollars of union funds by men like Dave Beck, James R. Hoffa, James Cross, and others is a grievous enough crime, but at best it is only a symptom of other more serious and underlying problems. A number of men in important union positions have come to regard unions as their own personal possessions.

Some fifteen years ago, a young and ambitious man battled his way to prominence in one of our nation's largest unions. He was a man of high ideals and dedicated to the cause of the men and women he represented in his union. As he rose in power within this union, his idealism faded. He lost touch with the very rank-and-file workers who had placed him in office. He joined country clubs, and his best friends became the employers with whom he negotiated contracts. Soon he began signing contracts with these employers weakening the working conditions and wages of his own union members. He started accepting loans from those with whom he was supposed to negotiate, and finally he started to steal from the union itself.

There is also the man who organized a small business in New York. His needs were modest and he treated his employees well. As the years progressed, his business grew. The size of his plant as well as his desire for profits increased. When a union official came to his door and said for a certain consideration he could sign a contract with him which did not even provide for the minimum wage of one dollar an hour, this man did not hesitate to make the payoff and sign the contract. By so doing, he participated in a fraud against several hundred employees, mostly uneducated Puerto Ricans who were unable to defend themselves, and reduced them to a form of economic slavery.

Neither of these cases is fictional. Both of these men appeared before the committee headed by Senator McClellan, and represent the type of testimony of which this committee has heard far too much . . .

Dangerous changes in American life are indicated by what is going on in America today. Disaster is our destiny unless we reinstall the toughness, the moral idealism which has guided this nation during its history. The paramount interest in oneself, for money, for material goods, for security, must be replaced by an interest in one another; an actual, not just a vocal interest in our country; a search for adventure, a willingness to fight, and a will to win; a desire to serve our community, our schools, our nation . . .

The 1960 Presidential Campaign

Robert Kennedy went to his father's house in Palm Beach, Florida, on April 1, 1959, to participate in the first meeting John Kennedy held to organize his entry into the presidential race. However, he was preoccupied with finishing not only the committee's work but also his manuscript for *The Enemy Within.* By the fall of that year, the Labor Reform Act was in place, the book was completed, and, after resigning from the committee, Kennedy became his brother's campaign manager. Then, for ten grueling months—through the Wisconsin and West Virginia primaries, the scramble for delegates leading up to a tumultuous Democratic convention in Los Angeles, and the campaign against Richard Nixon—he saw to it that the campaign stayed on track. He was tireless, hard driving, and, whatever else, effective. He did what had to be done, and, as he told Hugh Sidey, a Washington reporter for *Time*, "I'm not running a popularity contest . . . It doesn't matter if they like me—somebody has to be able to say no." Wherever he was, when he could, he made time to exhort campaign workers or speak to potential voters in small or large groups. Typical was one speech that he gave at a Democratic rally in Boston.

New England Democratic Meeting

BOSTON, MASSACHUSETTS

AUGUST 2, 1960

At the outset of this meeting of Democratic Party leaders of New England, we should discard any trace of the false notion that Senator Kennedy is assured of victory in his native region merely because he is one of its sons.

The political truths of the last three presidential elections must guide our thinking. In 1956 the Republican presidential candidate swept New England, repeating his accomplishment of 1952 . . . If the burden of overconfidence should rest anywhere, it should be on the shoulders of the opposition. Wisely, voters will not respond to residence, by itself, to reverse this trend.

The Democratic Party, rather, presents its case on a strong platform and strong candidates, on a proven record of past accomplishments, on the evidence that it alone has the vision and boldness to meet the challenge of leadership of the free world. The Democratic candidates for the presidency and vice presidency are men whose records of personal sacrifice and energy, decisiveness and experience, appeal to all citizens everywhere. And the American people this year will demand top quality in their leaders.

Democrats must campaign on the assumption the opposition will work hard. We must be more vigorous. We must provide the opportunity for each person who wants to take any active part in this most crucial election to assert himself on behalf of our mutual cause.

We must welcome to our ranks the disenchanted Republicans and resurgent Independents who have come to understand that a charming personality does not humanize or energize a stagnant political mechanism.

And we must stimulate the thousands of fellow citizens who fail to exercise their privilege to vote—the ultimate guarantee of a free society.

We will have to overcome the skillful propaganda of a Republican dynasty that piously repudiates Republicanism in action and adopts a blurred carbon copy of the platform Democrats have been advocating since the days of Wilson and Roosevelt.

We must unmask the Republican standard-bearer who can blithely ignore the failures of policy and action he has embraced for eight years. Our task will not be easy.

The Democratic Party has one great challenge—to inform the public of the truth. It is our greatest weapon. The people shall learn of the truth from us. And we have every confidence the best-informed citizens in the history of the United States will return a judgment in November that will be right.

John Kennedy narrowly was elected president, at forty-three the second youngest in the nation's history. Kennedy drew only 120,000 more popular votes than Nixon, out of nearly 69 million cast, but he finished comfortably ahead in the electoral vote, 303 to 219. The sibling relationship between John and Robert Kennedy was rare, bonded by strong family ties and enormous confidence in and respect for each other. John was an attractive, able candidate; Robert, an incredibly hard-working, action-oriented manager. "We simply had to run and fight," he told the writer Theodore H. White. "If we'd done one bit less of anything, then we might have lost."

PART TWO

Mr. Attorney General

January 20, 1961–
September 3, 1964

Introduction

Kennedy had serious misgivings about becoming his brother's attorney general. He believed that since the attorney general would be the federal point man in the rapidly escalating civil rights struggle, he would become an albatross around the president's neck; that the Kennedy name would inevitably create "tremendous political difficulties and problems for him—not only in the '64 election but in attempting to obtain the passage of legislation in other fields." Indeed, the potential for controversy was the reason Senator Abraham Ribicoff of Connecticut had turned down JFK's offer of the post. He said he didn't think a Jew should be putting black children in white Protestant schools in the South at the instruction of a Roman Catholic.

Robert Kennedy was, in fact, harboring thoughts of

striking out on his own—possibly as a college president or seeking political office in Massachusetts—but was persuaded by his father as well as the president-elect to become attorney general. The announcement aroused caustic disapproval in the press, notably in the *New York Times,* and in some legal and labor circles on the grounds of nepotism and that Kennedy's narrow legal experience had been confined to the Justice Department and Congress. He had never represented a client of any kind in a court proceeding.

The criticism did not make a dent on Capitol Hill. The Senate Judiciary Committee unanimously favored his confirmation, and the Senate confirmed him with only one negative vote. Nevertheless, Kennedy was mindful of the criticism, and though his arrival at Justice turned the spotlight on the department and transformed what had been a sleepy press office into a beehive of activity handling press inquiries, he was determined to keep a low profile while getting a grip on his job and organizing his staff. Accordingly, eleven weeks passed before he held a press conference. He made his first speech as the nation's sixty-fourth attorney general on April 21 to the American Society of Newspaper Editors, meeting in Washington, D.C. His first speech outside of the capital was on May 6, on civil rights, in Athens, Georgia.

Law Day

UNIVERSITY OF GEORGIA LAW SCHOOL

ATHENS, GEORGIA

MAY 6, 1961

In March, a young man named Jay Cox, president of the University of Georgia Law School's Student Advisory Council, arrived in Washington to invite Kennedy to deliver the school's Law Day speech. Two months earlier, on January 11, amid a riotous protest against the admission of two black students—Charlayne Hunter and Hamilton Holmes—Georgia officials had suspended the two "for their own safety." A federal court ordered them to be reinstated. Cox said the campus was tense and that Hunter and Hamilton were the objects of abuse almost daily as they attended classes. He warned there was risk that an appearance by the federal official most responsible for handling civil rights matters might make conditions worse. However, Cox said, he had come because he believed it could help the university move beyond the incident, that many Georgians would appreciate the gesture so soon after the violence.

Kennedy conferred with several southern politicians and friends, including Ralph McGill, publisher of the *Atlanta Constitution.* A few thought such an appearance would be unwise, but most agreed with Cox. The challenge of being the first top-ranking federal official to go to the Deep South to deliver a major civil rights speech and the opportunity of possibly easing tension on the campus weighted his decision to accept, but there was a third reason, one quite typical of Kennedy: He took an immediate liking to Cox, a young Georgian with a winning personality. Over the next five weeks, Kennedy and his staff honed the speech through seven rewrites. As would become his practice thereafter, he outlined to an aide the points he wanted to cover. For the Georgia speech it was his executive assistant, John Seigenthaler, who had been a reporter on the Nashville *Tennessean,* who wrote the first draft. John Bartlow Martin, a writer and JFK's ambassador to the Dominican Republic, sent another draft. Ralph McGill, James M. Landis (former dean of Harvard Law School), Sylvan Meyer (editor of the Gainesville, Georgia, *Times),* assistant attorneys general Burke Marshall (Civil Rights Division) and Nicholas Katzenbach (Office of Legal Counsel), and others all contributed ideas and comments. But the final version was more Kennedy's than anyone else's. He edited each draft, correcting sentences and rewriting whole paragraphs, until every line conveyed his views precisely.

There was a touch of summer humidity in the warm

spring morning when he landed in Athens. The night before, police had arrested five persons for painting KENNEDY GO HOME and YANKEE GO HOME on downtown sidewalks. Cox and a few fellow law students were at the airport, and they drove Kennedy to the university through quiet streets. One of the students asked why he had decided to speak on civil rights.

"It would have been hypocritical of me to come here and not do so," he replied.

For the first time since becoming attorney general, over three months ago, I am making something approaching a formal speech and I am proud that it is in Georgia . . . They have told me that when you speak in Georgia you should try to tie yourself to Georgia and the South, and even better, claim some Georgia kinfolk. There are a lot of Kennedys in Georgia, but as far as I can tell, I have no relatives here and no direct ties to Georgia, except one. This state gave my brother the biggest percentage majority of any state in the union, and in this last election that was even better than kinfolk.

We meet at this great university, in this old state, the fourth of the original thirteen, to observe Law Day.

In his proclamation urging us to observe this day,

the president emphasized . . . that to remain free, people must "cherish their freedoms, understand the responsibilities they entail, and nurture the will to preserve them." He then went on to point out that "law is the strongest link between man and freedom."

I wonder in how many countries of the world people think of law as the "link between man and freedom." We know that in many, law is the instrument of tyranny, and people think of law as little more than the will of the state or the party—not of the people.

And we know too that throughout the long history of mankind, man has had to struggle to create a system of law and of government in which fundamental freedoms would be linked with the enforcement of justice. We know that we cannot live together without rules which tell us what is right and what is wrong, what is permitted and what is prohibited. We know that it is law which enables man to live together, that creates order out of chaos. We know that the law is the glue that holds civilization together.

And we know that if one man's rights are denied, the rights of all are endangered. In our country the courts have a most important role in safeguarding these rights. The decisions of the courts, however much we might disagree with them, in the final analysis must be followed and respected. If we disagree with a court decision and thereafter irresponsibly assail the court and defy its rulings, we challenge the foundations of our society . . .

Respect for the law—in essence that is the meaning of Law Day—and every day must be Law Day or else our society will collapse . . .

Let me speak to you about three major areas of difficulty within the purview of my responsibilities that sap our national strength, that weaken our people, that require our immediate attention.

In too many major communities of our country, organized crime has become big business . . . Tolerating organized crime promotes the cheap philosophy that everything is a racket . . .

It is not the gangster himself who is of concern. It is what he is doing to our cities, our communities, our moral fiber . . .

The third area is the one that affects us all the most directly—civil rights. The hardest problems of all in law enforcement are those involving a conflict of law and local customs. History has recorded many occasions when the moral sense of a nation produced judicial decisions, such as the 1954 decision in *Brown v. Board of Education*, which required difficult local adjustments.

I have many friends in the United States Senate who are southerners. If these southern friends of mine are representative southerners—and I believe they are—I do not pretend that they believe with me on everything or that I agree with them on everything. But, knowing them as I do, I am convinced of this:

Southerners have a special respect for candor and plain talk. They certainly don't like hypocrisy. So, in discussing this third major problem, I must tell you candidly what our policies are going to be in the field of civil rights and why.

First let me say this: The time has long since arrived when loyal Americans must measure the impact of their actions beyond the limits of their own towns or states. For instance, we must be quite aware of the fact that 50 percent of the countries in the United Nations are not white; that around the world, in Africa, South America, and Asia, people whose skins are a different color than ours are on the move to gain their measure of freedom and liberty . . . And these people will decide not only their future but how the cause of freedom fares in the world.

In the United Nations we are striving to establish a rule of law instead of a rule of force. In that forum and elsewhere around the world our deeds will speak for us . . .

When parents send their children to school this fall in Atlanta, peaceably and in accordance with the rule of law, barefoot Burmese and Congolese will see before their eyes Americans living by the rule of law.

The conflict of views over the original decision in 1954 and our recent move in Prince Edward County [Virginia] is understandable. The decision in 1954 required action of the most difficult, delicate, and complex natures, going to the heart of southern institutions . . .

It is now being said that the Department of Justice is attempting to close all public schools in Virginia because of the Prince Edward situation.[1] That is not true, nor is the Prince Edward suit a threat to local control.

We are maintaining the orders of the courts. We are doing nothing more or less. And if any one of you were in my position you would do likewise, for it would be required by your oath of office. You might not want to do it, you might not like to do it, but you would do it. For I cannot believe that anyone can support a principle which prevents more than a thousand of our children in one county from attending public school—especially when this step was taken to circumvent the orders of the court.

Our position is quite clear. We are upholding the law . . .

In this case—in all cases—I say to you today that if the orders of the court are circumvented, the Department of Justice will act.

We will not stand by or be aloof. We will move.

Here on this campus, not half a year ago, you endured a difficult ordeal. And when your moment of truth came, the voices crying "force" were overridden by the voices pleading for reason.

And for this, I pay my respects to your governor, your legislature, and most particularly to you, the students and faculty of the University of Georgia. And I say, you are the wave of the future—not those who cry panic. For the country's future you will and must prevail.

I happen to believe that the 1954 decision was right. But my belief does not matter—it is the law. Some of you may believe the decision was wrong. That does not matter. It is the law. And we both respect the law. By facing this problem honorably, you have shown to all the world that we Americans are moving forward together—solving this problem—under the rule of law.

An integral part of all this is that we make a total effort to guarantee the ballot to every American of voting age—in the North as well as in the South. The right to vote is the easiest of all rights to grant. The spirit of our democracy, the letter of our Constitution and our laws require that there be no further delay in the achievement of full freedom to vote for all. Our system depends upon the fullest participation of all its citizens.

The problem between the white and colored people is a problem for all sections of the United States. And as I have said, I believe there has been a great deal of hypocrisy in dealing with it. In fact, I found when I came to the Department of Justice that I need look no further to find evidence of this.

I found that very few Negroes were employed above a custodial level. There were 950 lawyers working in the Department of Justice in Washington and only ten of them were Negroes. At the same moment the lawyers of the Department of Justice were bringing legal action to end discrimination, that same discrimination was being practiced within the department itself . . .

The federal government is taking steps to correct this.

Financial leaders from the East who deplore discrimination in the South belong to institutions where no Negroes or Jews are allowed, and their children attend private schools where no Negro students are enrolled. Union officials criticize southern leaders and yet practice discrimination within their unions. Government officials belong to private clubs in Washington where Negroes, including ambassadors, are not welcomed even at mealtime.

My firm belief is that if we are to make progress in this area—if we are to be truly great as a nation—then we must make sure that nobody is denied an opportunity because of race, creed, or color. We pledge, by example, to take action in our own backyard—the Department of Justice—we pledge to move to protect the integrity of the courts in the administration of justice. In all this, we ask your help—we need your assistance.

I come to you today and I shall come to you in the years ahead to advocate reason and the rule of law.

It is in this spirit that since taking office I have conferred many times with responsible public officials and civil leaders in the South on specific situations. I shall continue to do so. I don't expect them always to agree with my view of what the law requires, but I believe they share my respect for the law. We are trying to achieve amicable, voluntary solutions without going to court. These discussions have ranged from voting and school cases to incidents of arrest which might lead to violence.

We have sought to be helpful to avert violence and to get voluntary compliance. When our investigations indicate there has been a violation of law, we have asked responsible officials to take steps themselves to correct the situation. In some instances this has happened. When it has not, we have had to take legal action.

These conversations have been devoid of bitterness or hate. They have been carried on with mutual respect, understanding, and goodwill. National unity is essential, and before taking any legal action we will, where appropriate, invite the southern leaders to make their views known in these cases.

We, the American people, must avoid another Little Rock or another New Orleans.[2] We cannot afford them. It is not only that such incidents do incalculable harm to the children involved and to the relations among people. It is not only that such convulsions seriously undermine respect for law and order and cause serious economic and moral damage. Such incidents hurt our country in the eyes of the world . . .

For on this generation of Americans falls the full burden of proving to the world that we really mean it when we say all men are created free and are equal before the law. All of us might wish at times that we lived in a more tranquil world, but we don't. And if our times are difficult and perplexing, so are they challenging and filled with opportunity.

To the South, perhaps more than any other section of the country, has been given the opportunity and the challenge and the responsibility of demonstrating America at its greatest—at its full potential of liberty under law.

You may ask, will we enforce the civil rights statutes? The answer is: Yes, we will . . .

We can and will do no less.

I hold a constitutional office of the United States government, and I shall perform the duty I have sworn to undertake—to enforce the law, in every field of law and every region.

We will not threaten, we will try to help. We will not persecute, we will prosecute.

We will not make or interpret the laws. We shall enforce them—vigorously, without regional bias or political slant.

All this we intend to do. But all the high rhetoric on Law Day about the noble mansion of the law, all the high-sounding speeches about liberty and justice, are meaningless unless people—you and I—breathe meaning and force into it. For our liberties depend upon our respect for the law.

On December 13, 1889, Henry W. Grady of Georgia said these words to an audience in my home state of Massachusetts:

> *This hour little needs the loyalty that is loyal to one section and yet holds the other in enduring suspicion and estrangement. Give us the broad and perfect loyalty that loves and trusts Georgia alike with Massachusetts—that knows no South, no North, no East, no West, but endears with equal*

and patriotic love every foot of our soil, every state of our union.

A mighty duty, sir, and a mighty inspiration impels every one of us tonight to lose in patriotic consecration whatever estranges, whatever divides. We, sir, are Americans—and we stand for human liberty!

Ten days later Mr. Grady was dead, but his words live today. We stand for human liberty.

The road ahead is full of difficulties and discomforts. But as for me, I welcome the challenge, I welcome the opportunity, and I pledge my best effort—all I have in material things and physical strength and spirit to see that freedom shall advance and that our children will grow old under the rule of law.

One of the university's two black students, Charlayne Hunter, with press credentials from the *Atlanta Daily World,* was the only nonwhite in the hall.[3] Some eighteen hundred students, parents, professors, townspeople, and dignitaries heard Kennedy out in silence and remained silent when he finished and sat down. After twenty seconds or so, some began clapping. Then all were applauding, and as many in the audience gave

Kennedy a standing ovation, he stood up twice and bowed self-consciously.

The speech received little attention outside of Georgia. And as Kennedy spoke, forces were in motion that would mock his plea that "We, the American people, must avoid another Little Rock or another New Orleans." Two days earlier, seven black and six white men and women of the Congress of Racial Equality (CORE) had left Washington on a bus for a Freedom Ride through the South. They and other Freedom Riders would be brutally attacked by Ku Klux Klansmen in Anniston, Birmingham, and Montgomery, Alabama. President Kennedy would have to federalize Alabama National Guard units to rescue the Reverend Martin Luther King Jr. and fifteen hundred blacks—rallying in support of the Freedom Riders—from a Montgomery church that a mob was threatening to burn.

Press Conference, Department of Justice

WASHINGTON, D.C.

APRIL 6, 1961

Kennedy went to the Justice Department aware that the civil rights struggle would test the administration's mettle, but he vastly underestimated how quickly it would turn violent and spread across the country. Like most white Americans, he was only dimly aware of the desperate plight of most blacks. But the Law Day speech in Georgia was evidence that Kennedy was learning fast on the job. Thereafter, the government's role in the turbulent civil rights struggle and Kennedy's personal commitment to equal rights, along with his concerns about organized crime and the Cold War, formed the pegs for most of his speeches as attorney general. A fourth theme was just beginning to emerge: youth crime.

On March 5, 1961, after participating in a televised panel discussion on the challenge of international Communism, Kennedy walked three miles from the CBS studio at East 65th Street and York Avenue to a street corner in Harlem, where John Gomez, a New York City youth worker, had arranged separate meetings that evening with three gangs: two black and one Italian.

At each, the first question the young people asked Kennedy was:

"What are you doing here?"

"I came because I'm interested," he answered. He asked them where they lived, what they did, and what they thought the government should do to help them. Hesitantly, thoughtfully, they answered: provide more jobs and more facilities for recreation. Nothing more.

He returned to Washington wondering how the administration could make a difference. He talked to more youth crime experts and met with more gang members. A month later, at his first press conference as attorney general, while revealing that he was sending to Congress eight bills to curb organized crime and racketeering, Kennedy also announced the formation of what became known as the Allen Committee, to find out how the federal government might be more effective in reducing juvenile delinquency and crime, to be coordinated by his aide Dave Hackett, with Kennedy as chairman and chief advocate.

National Committee for Children and Youth Conference on Unemployed, Out-of-School Youth in Urban Areas

WASHINGTON, D.C.

MAY 24, 1961

Many people have asked me what connection the Department of Justice has with a conference on unemployment concerning out-of-school youths in urban areas. You may be wondering too. The reason is: Today, the Department of Justice, through the Bureau of Prisons, is providing institutional care and treatment for more than five thousand juvenile and youthful offenders. In fact, the Bureau of Prisons is now devoting nine of its thirty-one institutions to the ever-increasing number of young men, twenty-two and under, who are being turned over to us from the federal courts.

And the connection between these young men and your discussions are very relevant because the majority, prior to their apprehension, were out of school and out of work. In a recent special study of 350 of these men, only two could be regarded as skilled workers and only 54 as semi-skilled. Almost without exception, the study showed that these young men failed in school.

Despite the fact that most claimed to have an eighth-grade education, 35 percent were found to be functionally illiterate.

Contrary to some opinion, these young men in the main had not failed in school because of their lack of learning ability. The intelligence levels of most of them were within the normal range. These youths are typical of the school dropouts about whom we are concerned. Most were problem cases from their early school years. They were truant more often than other students. Their parents were not concerned about their academic achievements and many of them moved from school to school without ever making a satisfactory adjustment. We are doing everything we can to prepare these young people for the labor market upon their release. We are seeking new methods to make the transition from prison to community life more effective.

Even so, it is becoming more and more evident that even substantially increasing the effectiveness of our control and treatment efforts would not solve the problem.

For example, some 70 percent of all the prisoners in our custody have been committed previously. It is essential that we continue to take all possible steps to enforce the law vigorously and that we keep hardened prisoners from contaminating other young people. We recognize also that many young offenders may be redirected toward constructive activities through successful treatment efforts. However, these programs deal with the end results of delinquency—not the sources. We must broaden our attack and focus as much energy in the future on prevention as we have on control and treatment in the past. We must find ways to prevent spending more money for more institutions for more juveniles. And, unless something is done, this is what will happen. I cannot believe we will allow ourselves to meet this challenge in this way.

The effects of widespread changes in the social and economic life of our society has had a tremendous impact on the unskilled and poorly trained youth living in the slum areas of our large cities. These are the young people who now contribute most to the growing ranks of juvenile delinquency and youth crime and ultimately to adult criminal careers.

Where aspirations outstrip opportunities, law-abiding society becomes the victim. Attitudes of contempt toward the law are forged in this crucible and form the inner core of the beliefs of organized adult crime. To cope with these sources of criminality requires a broad concentrated effort to narrow the gap between legitimate aspirations of our deprived youths and the opportunities available to them. It means motivating and creating access to resources which will provide them opportunities to prepare for and pursue productive law-abiding careers.

Such a task exceeds the unaided capacities of individual families or even local communities. It requires a coordinated approach between different levels of government. The problem of delinquency today, we believe, is a problem of employment and education opportunities for preparation and achievement, as well as a problem of moral discipline and control. The failure to motivate and train *all* of our youths contributes to the growing problem of delinquency.

We cannot afford to ignore these signs of trouble. We cannot afford to ignore the rising social and economic costs. Instead, we must deal with the sources of the problem . . . The efforts of law enforcement and treatment agencies must be developed as a vital part of a large-scale preventive effort to channel the enterprise of our youths in all sectors of our society into equally useful and constructive activities. And this, with your help we must do.

Over the next three years, the committee and Hackett, working on a small budget, assisted local programs across the country and gathered data that formed the basis for part of President Lyndon Johnson's War on Poverty, particularly the community action concept. The committee was the first federal agency to work directly with indigenous leaders in poverty-stricken neighborhoods. In early 1963, it reported that, in effect, there were two criminal justice systems, one for the rich and one for the poor, most glaringly different in the matter of bail. Thousands of poor people were being kept in jail for weeks or even months after being arrested because they could not afford the cost of release.

"They may be innocent," Kennedy said in 1964. "They may be no more likely to flee than you or I. But they must stay in jail because, bluntly, they cannot afford to pay for their freedom." The committee recommended that federal prisoners be released on recognizance whenever possible, and shortly thereafter Kennedy instructed all United States attorneys to do so. Ultimately, the Justice Department drafted and Congress enacted the Criminal Justice Act of 1964, which provided indigent defendants with competent defense at every stage of proceedings against them; and the federal Bail Reform Act, which revised bail practices to assure that all persons, regardless of their financial status, would not be jailed needlessly pending their appearance to answer charges, to testify, or pending appeal, when detention served neither justice nor the public interest.

In recent years, it has again become clear that the American justice system does not mete out equal justice for indigent defendants. Funding cutbacks at all levels of government have been particularly harsh on public defenders, and the U.S. Supreme Court has to date failed to consider misuse of the bail process to incarcerate the poor who do not present flight or public safety concerns. Robert F. Kennedy Human Rights has been particularly active on these issues.

Two Speeches Made During a Worldwide Goodwill Tour

In February 1962, Robert and Ethel Kennedy flew around the world on a goodwill trip initially conceived as a three-country itinerary (Japan, Indonesia, and West Germany) that became a twenty-six day marathon journey, taking them to Japan, Taiwan, Hong Kong, Singapore, Indonesia, Vietnam, Thailand, Italy, West Germany, the Netherlands, and France, with refueling stops in Calcutta, Karachi, and Beirut on the fifteen-hour flight from Bangkok to Rome. A longtime friend of the Kennedy family, Dr. Gunji Hosono, director of the Japan Institute of International Affairs, had been urging Robert Kennedy for months to visit Japan. In the wake of violent anti-American riots that had caused President Eisenhower to cancel a trip to the country in 1960, Hosono believed that it was essential for the Japanese to know about the New Frontier. Secretary of State Dean Rusk advised Kennedy to accept Hosono's invitation and suggested that he also go to Indonesia because few U.S. officials had been there and a number of Communist country leaders, including Khrushchev, had made extended visits there. And in the fall of 1961, West Berlin Mayor Willy Brandt had invited Kennedy to give the Ernst Reuter Memorial Lecture at the Free University of Berlin. The State Department added the other stops on the trip.

Kennedy met with the top officials in each country but spent the major share of his time holding vigorous, frank discussions with students, many of whom lauded Marxism and denounced the United States.

On the trip, Kennedy made formal speeches at universities in Japan, Indonesia, West Berlin, and Bonn, always to capacity audiences, including 180,000 people who overflowed City Hall Square in West Berlin on a bitterly cold afternoon. In each speech, he contrasted democratic rights and freedoms with Communist rigidity and dogma. When he discussed current U.S. problems, accomplishments, and aspirations, Kennedy routinely provided historical perspectives, because, as he told the president on his return, he found that young people especially had serious questions about "our country and our way of life" and he wanted them to know what the United States had stood for in the past as well as today. Following are excerpts from two of those speeches.

Nihon University

TOKYO, JAPAN

FEBRUARY 6, 1962

Young men and women of my age in all countries, if they share nothing else, share the problems that torment our world and thus share responsibilities for all our futures. Our generation was born during the turmoil following the First World War. That war marked the dividing line—at least for the Western World—between the comfortable security of the nineteenth century and the instability and flux of our own time. After 1914, the world as the West knew it began to go to pieces. The old certitudes started to crumble away. World War II came as even a greater disaster.

Many young men, in my country as in yours, came out of the misery and chaos determined to do everything they could to spare the world another such catastrophe—and to lay the foundations for peace and social progress. This determination committed them to a public career—a career in politics or in government service. The president thus began his public life by running for Congress in 1946. I was a student at college then, and my fellow classmates and I worked hard in that campaign. My brother went on to the United States Senate at the age of thirty-five, and when he was elected president of the United States, although to you and me he was quite old, he was still a comparatively young man of forty-three!

In this, the United States gave recognition to and conferred responsibility upon the generation which was born in the First World War and raised in the Depression; which fought in the Second World War and launched its public career in the age of space.

I said that this generation grew up in an age of instability and flux. From one viewpoint, this is the worst of times in which to live—a time of anxiety and doubt and danger. But from another viewpoint, it is a time of great stimulation and challenge. It is a time of motion, when society is cutting away from old moorings and entering new historic epochs. It is a time that offers opportunity for initiative and ingenuity. It is a time that transforms life from routine into an adventure. The adventure of change may be a tragic adventure for many—a sad uprooting of cherished customs and institutions. Yet change is the one constant of history.

It certainly has been the dominating fact in the development of my own country. From the first moment of independence, the United States has been dedicated to innovation as a way of government and a way of life.

Not a decade has gone by in our nation's history in which we did not undergo new experiences and seek new challenges. We were born in a revolution against colonialism, and we have been dedicated ever since to revolution for freedom and progress.

My country has not been alone in pursuing these aims, and like all countries, the United States has made its share of mistakes. But at its best and at its most characteristic, the United States has been, above all, a progressive nation—a nation dedicated to the enlargement of opportunity for those President Andrew Jackson described as the humble members of society—the farmers, mechanics, and laborers.

The United States is a nation dedicated to the emancipation of women, the education of children, and above all to the dignity of the individual. This commitment to "life, liberty, and the pursuit of happiness" has inspired the essential motive of our national life: the unceasing search for new frontiers, not only frontiers of geography but also frontiers of science and technology and social and political invention and human freedom. These are the new frontiers which must be challenged and conquered by our generation—yours and mine. We must meet these problems and still maintain our dedication to democracy and freedom. To do so we must be imaginative and creative—not blindly wedded to the past . . . We in my country are by disposition and inheritance a people mistrustful of absolute doctrines and ideologies, persuaded that reason and experiment are the means by which free people fulfill their purposes. Yet we live in a century obsessed with ideology—a century that has been filled with leaders persuaded that they knew the secrets of history, that they were the possessors of absolute truth, and that all must do as they say or perish.

One of the great creative statesmen of our age was Franklin Roosevelt. He was creative precisely because he preferred experiment to ideology. He and the men of his time insisted that the resources of the democratic system were greater than many believed—that it was possible to work for economic security within a framework of freedom. Roosevelt and the Americans of his generation created a new society, far different from the unregulated and brutal economic order of the nineteenth century, the order on which Marx based his theories of capitalism.

This new society, since developed and molded by leaders of both of our political parties, is still loyal to its original revolutionary concept of the importance of the individual. Now under President Kennedy it sees as its goal service to mankind in ways never imagined years ago. It reaches out to protect us in our old age, it provides our youth with an ever better education, it seeks to keep our stock exchanges free from fraud and manipulation, and more and more it reaches out to newer and greater frontiers that will provide spiritually and economically a richer life. This is not the society condemned a hundred years ago as an era of unregulated capitalism based on laissez-faire . . .

We in the United States still have enormous problems which we have not solved, as for instance in the fields of civil rights, unemployment, and automation.

We make mistakes, our government and our people—internally as well as in our dealings with other nations. On some occasions we may move clumsily. However, to err is human and to move awkwardly is sometimes inevitable. The important thing is that as a government and as a people we fight and strive to make progress—to maintain freedom and to uphold at all times the dignity and independence of the individual.

. . . The resources of the earth and the ingenuity of man can provide abundance for all—so long as we are prepared to recognize the diversity of mankind and the variety of ways in which peoples will seek national fulfillment. This is our vision of the world—a diversity of states, each developing according to its own traditions and its own genius, each solving its economic and political problems in its own manner, all bound together by a respect for the rights of others, by a loyalty to the world community, and by a faith in the dignity and responsibility of man.

We have no intention of trying to remake the world in our image, but we have no intention either of permitting any other state to remake the world in its image . . .

In the unending battle between diversity and dogmatism, between tolerance and tyranny, let no one mistake the American position. We deeply believe that humanity is on the verge of an age of greatness—and we do not propose to let the possibilities of that greatness be overwhelmed by those who would lock us all into the narrow cavern of a dark and rigid system. We will defend our faith by affirmation, by argument, if necessary, and—heaven forbid that it should become necessary—by arms. It is our willingness to die for our ideals that makes it possible for those ideals to live . . .

Freedom means not only the opportunity to know but the will to know. That will can make for understanding and tolerance, and to ultimately friendship and peace.

The future stretches ahead beyond the horizon. No mortal man can know the answers to the questions which assail us today. But I am not ashamed to say that we in America approach the future, not with fear, but with faith—that we call to the young men and women of all nations to join us in a concerted attack on the evils which have so long beset mankind—poverty, illness, illiteracy, intolerance, oppression, war. These are the central enemies of our age—and I say to you that these enemies can be overcome.

Let us therefore pledge our minds and our hearts to this task, confident that though the struggle will take many generations, we shall be able to look back to this era as one in which our generations—joined as brothers—met our responsibilities and furthered the cause of peace at home and around the globe.

University of Indonesia

JAKARTA, INDONESIA

FEBRUARY 14, 1962

The outstanding spirit abroad in the world today is nationalism—nationalism closely linked with anti-colonialism.

Nationalism itself, of course, is nothing new. This self-determination performed the essential function of giving people an identity with their country and with each other. It became in some societies not merely an article of faith and common aspiration—but also a badge of conquest.

This was true of the nationalism which characterized the old Roman Empire, and was the driving force behind the German, Italian, and Japanese dictatorships in the days before World War II. It has not been true of our American nationalism. Nor is it true of the new nationalism loose in the world today. This nationalism has taken the form of "nonconquest," of disengagement from former economic and political ties. It is recreating in many parts of the world a sense of identity and of national aims and aspirations that the old order too often had sought to stifle.

The United States has always been sympathetic with this kind of national aspiration . . . The American people fought for national independence in 1776 and have been its spearhead ever since.

We know from our own history that the creation of a nation is not an easy matter. It took us years after our revolution—after we ceased to be a colony—to forge our thirteen original states into a nation that could act effectively and protect and promote the welfare of all its people . . . It is from our knowledge of difficulties we have faced, as well as from our dedication to the ideal of independence, that we have sought to aid new nations with technical and financial assistance during their crucial early years. Our aim is that they survive, develop, and remain proud and independent . . .

The United States has no desire to impose our conception of the role other nations should be allowed to assume. And I can tell you quite frankly we have no intention of permitting any other nation to enforce its system on other nations of the world.

On the contrary, the answer we have given and shall continue to give calls, as President Kennedy said, for the association of nations on a world and on a regional basis to defend the rights of the least on behalf of the whole . . .

But anti-colonialism is nothing if it does not follow national paths and remain true to its basic principle. If anti-colonialism is the struggle for freedom, then the new nations must remain free . . . Your president, in striving for an increase of living standards, has set as his goal "a just and prosperous society."

Its attainment, even to a moderate degree, is difficult. Men may differ as to what form of government will do the best job. Even within independent governments such as yours and mine, there is no rigid formula upon which all of us can agree . . .

But the important thing is, regardless of our differences, that we all hold firmly to the belief that a just and prosperous society is possible of achievement.

History has proved that a prosperous society, bringing decent living standards, education, and adequate medical care to all its citizens—not to a favored few—is made more nearly possible by men who remain dedicated to the principles of freedom than by those who are bound by a totalitarian system or an economic and political manifesto drafted a century ago as a solution for conditions which scarcely exist today . . .

In any society there are improvements that can be made, problems that remain unsolved. We have not fully attained for ourselves the prosperity that we seek for our people and all mankind. We can do better and we intend to do so. Our struggles against racial discrimination, our continuing struggle against want, our efforts to lift the levels of education so that any man may choose freely in the light of knowledge, all these struggles will continue. But we are making progress with them. That is what is important. We will not accept the status quo . . .

He returned convinced that the United States had to make a greater effort to win public confidence abroad, particularly of young people.

In his post-trip memo to his brother, Kennedy wrote: "It is apparent—in Indonesia, in Japan, and elsewhere—that the Communists have created the impression, not only in Asia, but across the world, that the young intellectual is for Communism and against our form of democratic society.

"The majority in fact is neither Communist nor pro-Communist. True enough, they are not pro–United States. They have serious questions about our country and our way of life. They frequently don't understand us. But with all of that, there is a tremendous reservoir of goodwill toward America and the American people."

He listed three suggestions:

1. Send groups of leading citizens abroad to talk about American history, philosophy, literature, and more practical matters with students, labor and farm leaders, and government officials.

2. See that federal information agencies did more to speak frankly about U.S. domestic problems and what the government and the American people were doing to deal with them.

3. Encourage other free countries to set up their own Peace Corps.

Kennedy elaborated his thoughts in his subsequent book *Just Friends and Brave Enemies,* in which he detailed the kinds of questions that were troubling the students in the lands he visited. The amount of misinformation as well as the lack of information about the United States and our system of government was "appalling." Expanding on his post-trip memo to the president, Kennedy said what needed to be done was for American officials and private citizens to go abroad to

> *discuss the United States frankly, admitting our shortcomings, stating our efforts to overcome them, and pointing out this country's accomplishments.*

> *We need people who have done their homework and who are not afraid to speak out . . . The history of America—and, in fact, America today—is full of men and ideas that are far more exciting and revolutionary than the systematic, pushbutton answers of Communist doctrine. I would like to see more Americans making this point. By not doing so, we are leaving most of the world with the illusion that the only modern philosophy belongs to Marx. Mere anti-Communism is not a philosophy; it is no substitute for really knowing what this country is about. Its history, its philosophy, its political thought are all available. They just need to be used.*

In the months that followed, he pressed people in and out of government to make time on foreign trips to speak to students and other groups, and many did.

Civil Rights, Foreign Policy, and Countering Communist Propaganda

In speaking during the spring and summer of 1962 to audiences that ranged from the Associated Press to the Virginia State Bar Association and to the American Booksellers Association, Kennedy often hammered on the need for Americans to do more overseas to counter Communist propaganda, and he motivated more than a few individuals and organizations to get involved.

But other matters preoccupied him. Since April 1961, when an invading brigade of CIA-trained, anti-Castro Cubans met disaster at Cuba's Bay of Pigs, he had become a key participant in the president's decision-making process dealing with the Soviet Union and other foreign policy matters. The drive against organized crime was picking up momentum. The administration was under fire from civil rights and civil liberties groups for appointing several segregationist jurists to federal judgeships in the South and for failing to press for civil rights legislation. Kennedy countered the criticism by citing progress in school desegregation; in eliminating discrimination in voting rights, transportation, and employment; and in protecting the integrity of court orders. Furthermore, though he viewed the plight of blacks in the cities as the most pressing urban problem, he was coming to see the civil rights struggle in a broader context: that of young men and women being denied opportunity, regardless of their race, religion, or ethnicity.

As the range of his vision expanded, so did his sense of urgency. Speaking about voting rights and education to the National Insurance Association in Los Angeles in July, he brought up the great French marshal [Louis-Hubert-Gonzalve] Lyautey [1854–1934], who "once asked his gardener to plant a tree. The gardener objected that the tree was slow-growing and would not reach maturity for one hundred years. The marshal replied, 'In that case, there's no time to lose. Plant it this afternoon.' Our youths have long lives ahead of them. Today an America of equal opportunity for all its citizens is just around the corner, and we have no time to lose. Let's plant our trees this evening." That urgency only grew, when two months later he returned to California.

On the Duty of Lawyers

UNIVERSITY OF SAN FRANCISCO

SEPTEMBER 29, 1962

On September 29, 1962, a mounting crisis at the all-white University of Mississippi over the court-ordered admission of James Meredith, a twenty-nine-year-old black air force veteran, compelled Kennedy to cancel a trip to speak that evening at the dedication of the University of San Francisco Law School's Kendrick Hall. For two weeks, Robert and President Kennedy, increasingly frustrated, had sought unsuccessfully to persuade Mississippi's segregationist governor, Ross Barnett, to cease resisting the court orders to enroll Meredith. Barnett based his opposition on the long-discredited, pre–Civil War doctrine of interposition: that state officials had the right to place state sovereignty between its people and the federal government. After Kennedy informed his hosts that he would not be at the dedication, his press aide Ed Guthman showed him a new draft of the speech criticizing the American Bar Association, as well as lawyers in Mississippi, for not speaking out against Barnett's defiance.

"There's nothing wrong with saying it," Kennedy said. "Let's get the speech out. I wouldn't have believed it could have happened in this country, but maybe now we can understand how Hitler took over Germany!"

The text of the speech was released to the press. It was filled with examples of attorneys who had demonstrated both physical and moral courage. These included Andrew Hamilton, the Philadelphia lawyer who in 1735 successfully defended printer John Peter Zenger against charges of seditious libel; John Adams, who was pilloried as a traitor when he defended British soldiers and an officer for firing on a crowd of Boston colonists in 1770; Clarence Darrow, who gave up his career as a corporate lawyer in 1895 to defend Socialist Eugene Debs and other officers of the American Railway Union against charges of criminal conspiracy, evolving out of the union's strike against the railroad for which Darrow had been counsel; and Harold Medina, a leading New York trial lawyer who, during World War II, defended, without pay, a German national charged with high treason. Kennedy went on:

All these attorneys rose above the interests of their pocketbooks. They were men who freely stepped across the boundary of their own legal specialty, often at the cost of their popularity. They served in a role which, throughout history, has challenged the finest of our lawyers—that is, the role of the citizen. By that I mean a great deal more than the right to vote, or to obtain a passport, or even to speak and worship freely.

Since the days of Greece and Rome when the word *citizen* was a title of honor, we have often seen more emphasis put on the right of citizenship than on its responsibilities. And today, as never before in the free world, responsibility is the greatest right of citizenship, and service is the greatest of freedom's privileges.

Lawyers have their duties as citizens but they also have special duty as lawyers. Their obligations go far deeper than earning a living as specialists in corporation or tax law. They have a continuing responsibility to uphold the fundamental principles of justice from which the law cannot depart . . .

One of my great disappointments in our present efforts to deal with the situation in Mississippi as lawyers has been the absence of an expression of support from the many distinguished lawyers of the state. I realize in that difficult social situation that to defend the fundamental principles of respect for the law and compliance with federal court orders would be unpopular and require great courage.

I also understand that many of them may not agree with the decision in *Brown v. Board of Education*, but whether they agree or not they still have their obligations as lawyers, and they have remained silent. However, I might also note that there have been no pronouncements in this matter by the American Bar Association.

The next day, Sunday, with a force of United States marshals and soldiers assembled in Memphis, sixty miles north of the University of Mississippi campus at Oxford, Barnett agreed to a plan whereby the marshals would escort Meredith onto the campus and the governor would announce that he had been outwitted and was "ceasing resistance." Kennedy hurriedly dispatched several of his aides, including Guthman and Deputy Attorney General Nicholas Katzenbach, to Mississippi. That evening, when Katzenbach, Guthman, and Deputy Assistant Attorney General John Doar, who had been on the scene for several weeks, and federal marshals brought Meredith to the campus, students rioted, augmented by more than one thousand hate-filled men, some of whom had driven long distances to "stand with Ross."

A furious fight ebbed and flowed through the night until military police and a Second Division battle group from Fort Benning, Georgia, restored order. A French newsman, Paul Guihard, and a bystander, Ray Gunter, an Oxford, Mississippi, repairman, were killed. One hundred sixty marshals—more than a third of the contingent—were injured, of whom twenty-eight were wounded by gunfire, and as many demonstrators. But at eight o'clock the next morning Meredith, guarded by federal marshals and soldiers, registered without incident. As Walter Lord noted in his book on the incident, *The Past That Would Not Die,* "Meredith ultimately graduated on a sunny, peaceful morning in August 1963 that offered a striking contrast to the wild, bloody night when he arrived."

Testimonial Dinner for John Reynolds

MILWAUKEE, WISCONSIN

OCTOBER 6, 1962

When Robert Kennedy first spoke in public after the University of Mississippi riot—at a dinner honoring John Reynolds, the Democratic candidate for governor of Wisconsin—rather than deploring the violence or taking credit for having upheld court orders, he lauded what individuals had done, affirming, as he often did in speeches, his belief that one person could make a difference regardless of the odds.

We are now in the midst of another political campaign.

Elections remind us not only of the rights but the responsibilities of citizenship in a democracy. Yet we live in a time when the individual's opportunity to meet his responsibilities appears circumscribed by impersonal powers beyond his influence. On the surface the individual in American society is pressed on all sides by the mightiest materialistic force in man's history. The power of atomic weapons seems to dwarf the heroism of any individual soldiers, and City Hall looms too big to fight in a hundred walks of life.

But even today there is so much that a single person can do with faith and courage, and we have had a number of outstanding examples just this week . . . James Meredith brought to a head and lent his name to another chapter in the mightiest internal struggle of our time. At the same time . . . there were five hundred United States marshals, most of them from the southern states, who remained true to their orders and instructions and stood with great bravery to prevent interference with federal court orders.

A troop of armored cavalry—men from Oxford, Mississippi—were the first soldiers to come to the aid of the marshals. Some of these young men had graduated only last June from the University of Mississippi. As one of them said: "We don't like being here, but we don't like that mob shooting at you either."

As the Mississippi crisis was ending, a crisis of frightful proportions was about to shake the world. On Monday, October 15, 1962, a U-2 spy plane brought back photographs that showed without doubt that the Soviets were deploying surface-to-surface nuclear missiles in Cuba. Robert Kennedy was informed the next morning in the Oval Office, and for the next thirteen days he was totally involved in the president's stressful decision-making process and intricate diplomacy to get the missiles out.

"In all this, Robert Kennedy was the indispensable partner," historian Arthur M. Schlesinger Jr. wrote in 1978:

> *Without him, John Kennedy would have found it far more difficult to overcome the demand for military action . . . It was Robert Kennedy who oversaw the Executive Committee [of the National Security Council], stopped the air-strike madness in its tracks, wrote the reply to the Khrushchev letter, conducted the secret negotiations with [Soviet Ambassador Anatoly] Dobrynin.*

The confrontation ended Sunday morning, October 28, when Khrushchev sent a message that the Soviets would dismantle and withdraw the missiles. That afternoon, though relieved and exhausted, Robert Kennedy decided he would keep a long-standing commitment to speak at an American Jewish Congress dinner in New York, where he was to receive the Rabbi Stephen S. Wise Award "for advancing human freedom." Donning a tuxedo, he flew to New York, and, near the end of his speech, spoke about the missile crisis.

We meet tonight in a time of grave crisis with our attention fixed on the waters of the Caribbean and the once peaceful hills and fields of Cuba.

The confrontation between the United States and the Soviet Union is in reality a confrontation of all people who believe in human dignity and freedom with those who believe the state is supreme. It is that fact, not the drama of the particular moment, which is of real significance.

In our society, laws are administered to protect and expand individual freedom, not to compel individuals to follow the logic other men impose on them. The tyranny of Communism is as old as the pharaohs and the pyramids—that the state stands above all men and their individual aspirations. And that is why we oppose it, because by force and subversion it seeks to impose its tyranny all around the world.

We will not win this struggle merely by confronting the enemy. What we do at home, in the final analysis, is just as important. Thus we all must accelerate our efforts to banish religious prejudice, racial discrimination, and any intolerance which denies to any Americans the rights guaranteed them by the Declaration of Independence and the Constitution.

That is what this crisis is all about; that is why our ships are on station in the Caribbean and why American soldiers are on duty tonight in West Berlin, South Vietnam, and South Korea. They are there for the same reason the Maccabees stood their ground against Antiochus—for human dignity and freedom.

It has been said that each generation must win its own struggle to be free. In our generation, thermonuclear war has made the risks of such struggles greater than ever. But the stakes are the same: the right to live in dignity according to the dictates of conscience and not according to the will of the state.

After that speech Robert Kennedy hardly mentioned the Cuban Missile Crisis in speaking publicly, although the Cold War was often an integral part of his remarks, along with civil rights and criminal justice. But it weighed on his mind, and he provided a detailed report and definitive comments in *Thirteen Days,* a book published in 1969 that he began in 1967 but did not finish. Hauntingly, the final chapter was to have discussed whether the United States or any government had the right to bring its people and the world to the brink of nuclear devastation.

The unfinished manuscript ended as follows:

The final lesson of the Cuban Missile Crisis is the importance of placing ourselves in the other country's shoes. During the crisis, President Kennedy spent more time trying to determine the effect of a particular course of action on Khrushchev or the Russians than on any other phase of what he was doing. What guided all his deliberations was an effort not to disgrace Khrushchev, not to humiliate the Soviet Union, not to have them feel they would have to escalate their response because their national security or national interests so committed them.

That is why he was so reluctant to stop and search a Russian ship; this was why he was so opposed to attacking the missile sites. The Russians, he felt, would have to react militarily to such actions on our part.

Thus the initial decision to impose a quarantine rather than to attack; our decision to permit the *Bucharest* to pass; our decision to board a non-Russian vessel first; all these and many more were taken with a view to putting pressure on the Soviet Union but not causing a public humiliation.

Miscalculation and misunderstanding and escalation on one side bring a counter-response. No action is taken against a powerful adversary in a vacuum. A government or people will fail to understand this only at their great peril. For that is how wars begin—wars that no one wants, no one intends, and no one wins.

Each decision that President Kennedy made kept this in mind. Always he asked himself: Can we be sure that Khrushchev understands what we feel to be our vital national interest? Has the Soviet Union had sufficient time to react soberly to a particular step we have taken? All action was judged against that standard—stopping a particular ship, sending low-flying planes, making a public statement.

President Kennedy understood that the Soviet Union did not want war, and they understood that we wished to avoid armed conflict. Thus, if hostilities were to come, it would be either because our national interests collided—which, because of their limited interests and our purposely limited objectives, seemed unlikely—or because of our failure or their failure to understand the other's objectives.

President Kennedy dedicated himself to making it clear to Khrushchev by word and deed—for both are important—that the U.S. had limited objectives and that we had no interest in accomplishing those objectives by adversely affecting the national security of the Soviet Union or by humiliating her.

Later, he was to say in his speech at American University in June of 1963: "Above all, while defending our own vital interests, nuclear powers must avert those confrontations which bring an adversary to the choice of either a humiliating defeat or a nuclear war."

During our crisis talks, he kept stressing the fact that we would indeed have war if we placed the Soviet Union in a position she believed would adversely affect her national security or such public humiliation that she lost the respect of her own people and countries around the globe. The missiles in Cuba, we felt, vitally concerned our national security, but not that of the Soviet Union.

This fact was ultimately recognized by Khrushchev, and this recognition, I believe, brought about his change in what, up to that time, had been a very adamant position. The president believed from the start that the Soviet chairman was a rational, intelligent man who, if given sufficient time and shown our determination, would alter his position. But there was always the chance of error, of mistake, miscalculation, or misunderstanding, and President Kennedy was committed to doing everything possible to lessen that chance on our side. The possibility of the destruction of mankind was always in his mind. Someone once said that World War Three would be fought with atomic weapons and the next war with sticks and stones . . .

Barbara Tuchman's *The Guns of August* had made a great impression on the president. "I am not going to follow a course which will allow anyone to write a comparable book about this time," he said to me that Saturday night, October 26. "If anybody is around to write after this, they are going to understand that we made every effort to find peace and every effort to give our adversary room to move. I am not going to push the Russians an inch beyond what is necessary."

After it was finished, he made no statement attempting to take credit for himself or for the administration for what had occurred. He instructed all members of the Ex Comm [Executive Committee of the National Security Council] and government that no interview should be given, no statement made, which would claim any kind of victory. He respected Khrushchev for properly determining what was in his own country's interest and what was in the interest of mankind. If it was a triumph, it was a triumph for the next generation and not for any particular government or people.

At the outbreak of the First World War the ex-chancellor of Germany, Prince von Bulow, said to his successor, "How did it all happen?" "Ah, if only we knew," was the reply.

Senate Commerce Committee

WASHINGTON, D.C.

JULY 1, 1963

In the winter and spring of 1963—the centennial of President Lincoln's signing of the Emancipation Proclamation freeing "all slaves in areas still in rebellion"—angry, often violent civil rights confrontations flared from the eastern shore of Maryland to Texas. The worst occurred in Birmingham, Alabama. Over several weeks, blacks, with many children in their ranks and led by the Reverend Martin Luther King Jr., marched daily to protest discrimination in employment and public accommodations. Each day police routed them, and across the nation people saw on the network news the marchers being beaten with nightsticks, knocked down by water from high-pressure hoses, and attacked by police dogs.

Ultimately, Burke Marshall, assistant attorney general in charge of the Civil Rights Division, negotiated an uneasy truce. On May 17, the day after Marshall returned to Washington, he and other Justice Department officials flew with Robert Kennedy to Asheville, North Carolina, where Kennedy spoke at the North Carolina Cold War Seminar. Aboard the plane they drafted the essential elements of what would become the Civil Rights Act of 1964.

At the University of Alabama on June 11, when two young blacks, Vivian Malone and James Hood, came to register for the summer semester, Alabama Governor George C. Wallace stood at the entrance to Foster Auditorium, fulfilling his campaign pledge to "stand in the schoolhouse door" to prevent school desegregation. However, with Deputy Attorney General Katzenbach on the scene, Wallace marched off without confronting the students, and they enrolled without violence. That night President Kennedy, in a televised report to the nation, said America faced a "moral issue" that was not confined to the South and that it was time for bold action.

"If an American, because his skin is dark, cannot eat lunch in a restaurant open to the public," he said, "if he cannot send his children to the best school available, if he cannot vote for the public officials who represent him, if, in short, he cannot enjoy the full and free life which all of us want, then who among us would be content to stand in his place? Who among us would then be content with counsels of patience and delay?"

Later that night in Jackson, Mississippi, Medgar Evers, state director of the National Association for the Advancement of Colored People, was shot in the back and killed by a white man as he stepped out of his car in his driveway.

On June 19, President Kennedy sent the administration's omnibus civil rights bill to Congress. It included proposals languishing in Congress to outlaw complicated tests and other unfair practices that prevented blacks from voting. It also proposed a ban on the use of federal funds in discriminatory state or local programs and, most controversial of all, a law to end discrimination in places of public accommodation—restaurants, stores, hotels, lunch counters, and theaters. Robert Kennedy led the administration's fight in Congress, appearing before the House Judiciary Committee on July 1, and the Senate Judiciary Committee on July 18 and August 23. On each occasion, he argued for the bill with a mass of evidence of discriminatory practices and with a powerful moral plea, typified by the following excerpt from his testimony before the Commerce Committee on July 1.

The law will set no precedent in the field of governmental regulation, nor will it unjustly infringe on the rights of any individual. The only right it will deny is the right to discriminate—to embarrass and humiliate millions of our citizens in the pursuit of their daily lives.

The places of business covered by this law are public in a very real sense: They are not private homes or clubs. They deal with the general public. They invite and in fact compete for public patronage.

Plainly, when a customer is turned away from such a place because of the color of his skin, it imposes a badge of inferiority on that citizen which he has every right to resent. And in addition to the insult, consider the physical and financial inconvenience suffered by Negroes through such discrimination.

A white person, traveling in a part of the country where discrimination is customary, can make reservations in advance or stop for food and lodging where and when he will.

For the Negro it is not so simple. If he makes reservations without first determining whether or not the establishment will accept people of his race, he may well find on his arrival that the reservation will not be honored—or that it will somehow have been mislaid. His alternative is to subject himself or his family to the humiliation of rejection at one establishment after another—until, as like as not, he is forced to accept accommodations of inferior quality, far removed from his route of travel.

White people of whatever kind—even prostitutes, narcotics pushers, Communists, or bank robbers—are welcome at establishments which will not admit certain of our federal judges, ambassadors, and countless members of our armed forces . . . If Congress can, and does, control the service of oleomargarine in every restaurant in the nation, surely it can insure our nonwhite citizens access to those restaurants. [He then listed thirty-eight federal statutes, based on the commerce clause of the Constitution, that regulate private businesses and property.]

So I think it only fair, Mr. Chairman, to declare that this bill does not seriously or significantly interfere with private property rights, nor does it extend any principle of federal regulation. Therefore, the argument that it does should be rejected as a smoke screen. The real issue is whether Congress should or should not ban racial discrimination in places open to the public . . .

Recent events . . . are vivid evidence that this bill is needed. There has been a good deal of talk decrying the Negro demonstrations. I say that any discussion of this problem which dwells solely on the demonstrations and not on the causes of those demonstrations is not going to solve anything.

This is the lesson of history. There still are workers and their children alive today in mining communities all through the Appalachian area and in industrial centers all over the country who remember the time before the Wagner Act, the wage and hour laws, and adequate mine safety and factory-inspection acts. In those days, men and women were striving to form unions or to attract legislative attention to their condition. Their meetings were suppressed. Their demonstrations were put down. Their pickets were arrested.

Repression on one side often produces violence on the other. The Haymarket Square Riot of 1886, the violence of the Little Steel Strike of 1938, and hundreds of similar incidents, large and small, are tragic cases in point.

But we learned from them and concentrated our attention upon substantive evils that gave rise to the outbreaks. That is what the Interstate Public Accommodations Act of 1963 attempts to do—to concentrate upon the evils which have caused the recent demonstrations.

With the adoption of the Thirteenth, Fourteenth, and Fifteenth Amendments, the American Negro was freed from slavery and made a citizen of full standing—on paper, at least. But for most of the past hundred years we have imposed the duties of citizenship on the Negro without allowing him to enjoy the benefits.

We have demanded that he obey the same laws as white men, pay the same taxes, fight and die in the same wars. Yet in nearly every part of the country, he remains the victim of humiliation and deprivation no white citizen would tolerate. All thinking Americans have grown increasingly aware that discrimination must stop—not only because it is legally insupportable, economically wasteful, and socially destructive, but above all because it is morally wrong . . .

The federal government has no moral choice but to take the initiative. How can we say to a Negro in Jackson, "When a war comes you will be an American citizen, but in the meantime, you're a citizen of Mississippi and we can't help you"? . . .

The United States is dominated by white people, politically and economically. The question is whether we, in this position of dominance, are going to have, not the charity, but the wisdom to stop penalizing our fellow citizens whose only fault or sin is that they were born.

A year later Congress passed the bill with its key provisions not only intact but slightly strengthened, and President Johnson signed it into law on July 2, 1964.

During the summer and fall of 1963, there was another important civil rights issue for which Robert Kennedy also rallied support in his public speeches. It was making sure that impoverished criminal defendants got the legal counseling that the United States Supreme

Court had ruled unanimously was their constitutional right on March 18, 1963, in the case of *Gideon v. Wainwright.* Two weeks later—just a week before President Kennedy was assassinated in Dallas—Robert Kennedy, speaking at a regional meeting of the American Bar Association in Cleveland, Ohio, said:

As the law now stands, court-appointed defense attorneys receive no pay, nor are they reimbursed for out-of-pocket expenses. They receive no investigative or expert help, and they are not appointed until long after the time of arrest—when witnesses often have vanished and leads grown stale. Moreover, those appointed all too often lack courtroom experience.

The result, borne out of several recent studies, is that poor defendants stand less chance of obtaining full justice than wealthy ones. Pleas of guilty are entered much more frequently by defendants with appointed counsel than by those with privately retained attorneys—and they have less chance of getting charges against them dismissed. If they go to trial, they have less chance of being acquitted; and if convicted, they have less chance of getting probation instead of jail.

This is highly disturbing when we realize that almost ten thousand persons charged with federal crimes each year—more than 30 percent of the total—have court-appointed counsel.

Then, in a typical Robert Kennedy challenge to his audience, he said:

Obviously, not all Americans accused of crimes are poor. Some, in fact, are extremely rich . . . I'd like to read a statement once written by a highly successful businessman. "This American system of ours—" he said, "call it Americanism, call it capitalism, call it what you like—gives each and every one of us a great opportunity if we only seize it with both hands and make the most of it."

It wasn't Henry Ford who wrote those patriotic words, and it wasn't Andrew Carnegie.

It was Al Capone.

He was speaking from a wealth of personal experience. The American system did indeed give Al Capone and his kind great opportunity—they became tycoons of crime, millionaire rulers of an industry that flourished in violation of the law.

. . . The key men in the big crime syndicates have enough money and enough power to turn the very machinery of judicial process to their own advantage. They can, and do, hire excellent legal brains to defend them—and to help ward off their apprehension in the first place. They can, and do, bring influence to bear on jurors, either by bullying or bribery. They can, and do, cause cases against them to collapse by their power to intimidate prosecution witnesses . . .

In the problems of organized crime, as well as in those of the poor defendant, federal legislation is pending that may improve our ability to see that justice is done. But legislation is not enough. A clear and definite obligation exists for all members of the legal profession . . .

Wherever organized crime exists—whether or not it involves political or police corruption—the lawyer should see it as his duty to call it to public attention, to work however he can to make his fellow citizens aware of the waste and menace they are inadvertently supporting . . .

In all his dealings—with rich and poor, with the innocent and the criminal—a good lawyer is an embodiment of the sanctity of the law, a living reminder that ours is a government of laws and not of men—or of money.

After Dallas (and JFK's Death)

After JFK's death on November 22, 1963, Kennedy did not appear before a large American audience until he spoke at a Friendly Sons of St. Patrick dinner in Scranton, Pennsylvania, nearly four months later. He went reluctantly and only after insistent urging by the powerful, canny Representative Dan Flood of Wilkes-Barre, for through the crushing weeks following JFK's death, he had been desolate: holding his grief inwardly, functioning tautly, and wondering bleakly what he would do with his life.

As spring approached he showed signs of perking up, but one often saw, just below the surface, signs of a new somberness and fatalism, caused not only by the trauma of the assassination but also by the realization that any decision he would make about his future would probably depend on events beyond his control.

Scranton marked a turning point.

When the Kennedy family's plane, the *Caroline,* landed at Wilkes-Barre Scranton Airport on an overcast afternoon, more than two thousand people, most of them young, broke through police lines and, jumping, shouting, and laughing, they crowded in so tightly trying to touch Kennedy that he could not move from the bottom of the ramp until police cleared a path.

He was taken to Scranton to officiate at groundbreaking ceremonies for the new John F. Kennedy Elementary School, and then he was driven to nearby Yatesville to speak briefly at a St. Patrick's Day dinner. Heavy, wet snow had begun to fall, and along the route about ten thousand men, women, and children huddled under umbrellas to catch a glimpse of him. They were still lining the road when he drove back to Scranton an hour later.

His appearance before the Friendly Sons at the Hotel Casey in Scranton was one of the most difficult he ever made. He had drafted a sentimental Irish speech, and for the ending he had written: "I like to think—as did President Kennedy—that the emerald thread runs into the cloth we weave today . . . and I like to think that his policies will survive and continue, as the cause of Irish freedom survived the death of the Liberator, Owen Roe O'Neill."

He planned to close with a poem written after O'Neill's death, when grief overwhelmed Ireland, which ended:

We're sheep without a shepherd
When the snow shuts out the sky—
Oh! Why did you leave us, Owen?
Why did you die?

Ed Guthman eliminated the poem while working into the speech some changes Kennedy wanted.

"Why did you do that?" Kennedy asked.

"Because you'll never get through it," Guthman said.

"I've been practicing in front of a mirror," Kennedy said. "I can't get through it yet—but I will!"

And in the hotel's grand ballroom he did—just barely; but among the stalwart sons of Erin in the audience, many a man wept openly.

Friendly Sons of St. Patrick of Lackawanna County

SCRANTON, PENNSYLVANIA

MARCH 17, 1964

I'm aware, of course, of the notable number of sons of St. Patrick who live here in Scranton, and as a son of St. Patrick myself, I know how friendly you've always been—to President Kennedy in everything he did—and to me whenever I've been here.

So I think of these things in addition to the bonds of common kinship that the Irish everywhere feel on St. Patrick's Day. This is the day, you know, when legend has it that three requests were granted to St. Patrick by an angel of the Lord, in order to bring happiness and hope to the Irish.

First, that on this day the weather should always be fair to allow the faithful to attend church. Second, that on every Thursday and Saturday, twelve Irish souls should be freed from the pains of hell. And, third, that no outlander should ever rule over Ireland.

Though I have not received the latest weather report from the Emerald Isle, I am confident that no rain fell there today—officially. Who pays heed to a little Irish mist?

And I have reason to believe that the twelve Irishmen have been regularly released from the nether regions as promised. [U.S. District] Judge [William J.] Nealon just told me he thinks that several of them are here tonight.

We need have no concern over the third promise; in Ireland they are celebrating this day in freedom and liberty. But you and I know that life was not always this good for the Irish, either back in the old country or here in America.

There was, for example, that black day in February 1847 when it was announced in the House of Commons that fifteen thousand people a day were dying of starvation in Ireland . . .

So the Irish left Ireland. Many of them came here to the United States. They left behind hearts and fields and a nation yearning to be free. It is no wonder that James Joyce described the Atlantic as a bowl of bitter tears, and an earlier poet wrote, "They are going, going, going, and we cannot bid them stay."

This country offered great advantages, even then. But no one familiar with the story of the Irish here would underrate the difficulties they faced after landing in the United States. As the first of the racial minorities, our forefathers were subject to every discrimination found wherever discrimination is known.

But many of them were gifted with boundless confidence that served them so well. One was a pugilist from my native Boston. John L. Sullivan won the heavyweight championship of the world not too many years after the flood tide of Irish emigration to this country, and in 1887 he toured the British Isles in triumph.

Some idea of Irish progress can be gathered from his cordial greeting to the Prince of Wales, later Edward VII. John L. said: "I'm proud to meet you. If you ever come to Boston be sure to look me up. I'll see that you're treated right."

And, referring to the Prince, he later added with Irish generosity "Anyone can see he's a gentleman. He's the kind of man you'd like to introduce to your family."

Irish progress here has continued. It was some time ago that the late Fred Allen defined the "lace curtain Irish" as those who "have fruit in the house when no one's sick."

But it was less than nine months ago when President Kennedy, in touring Ireland, used to ask the crowds he talked with how many had cousins in America. The usual response was for nearly every hand in the crowd to be raised. It was with great delight that he was able to reply: "I've seen them, and they're doing well." And so, it is my great delight to be with you here tonight as we take a few moments to share the rich heritage of the Irish.

It's worth noting, I think, that all the wealth of our legacy stems from a small island in the far Atlantic with a population one quarter the size of the state of Pennsylvania.

The Irish have survived persecution in their own land and discrimination in ours. They have emerged from the shadow of subjugation into the sunlight of personal liberty and national independence. And they have shared the struggles for freedom of more than a score of nations across the globe . . .

Indeed, Ireland's chief export has been neither potatoes nor linen but exiles and immigrants who have fought with sword and pen for freedom around the earth . . .

And other Irishmen in other years, going into battle with the Union Army—a green sprig in their hats—bore the brunt of the hopeless assaults on the Confederate heights at Fredericksburg. Twelve hundred soldiers of the Irish Brigade went into action that bitterly cold December day in 1862. Only 280 survived, as President Kennedy noted last summer when he presented the brigade's battle flag to the Irish people. "Never were men so brave," General Robert E. Lee said of the Irish Brigade.

And of themselves, the Irish soldiers said:

War-battered dogs are we,
Gnawing a naked bone;

Fighters in every land and clime
For every cause but our own.

Today the Irish enjoy their freedom at a time when millions of people live in deprivation and despair under totalitarian dictatorships stretching eastward from the Wall in Berlin to the troubled borders of South Vietnam . . .

So the first point I'd like to make arises from the traditional Irish concern for freedom—everywhere. I know of few in our land—and I hope none in this room—who would ignore threats to peace and freedom in far-off places. We realized, as John Boyle O'Reilly once wrote, that:

The world is large, when its weary leagues
Two loving hearts divide;
But the world is small, when your enemy
Is loose on the other side.

No problem weighs heavier on the conscience of free men than the fate of millions in iron captivity. But what is taking place on the other side of the Iron Curtain should not be the only matter of concern to us who are committed to freedom. I would hope that none here would ignore the current struggle of some of our fellow citizens right here in the United States for their measure of freedom. In considering this, it may be helpful for us to recall some other conditions that existed in Ireland from 1691 until well in the nineteenth century again, which our forefathers fought.

We might remember, for instance, that in Ireland of 1691 no Irish Catholic could vote, serve on a jury, enter a university, become a lawyer, work for the government, or marry a Protestant. And our pride in the progress of the Irish is chilled by the tragic irony that it has not been progress for everyone.

We know that it has not been progress for humanity. I know because so much work of the Department of Justice today is devoted to securing these or comparable rights for all Americans in the United States in 1964.

There are Americans who—as the Irish did—still face discrimination in employment—sometimes open, sometimes hidden. There are cities in America today that are torn with strife over whether the Negro should be allowed to drive a garbage truck; and there are walls of silent conspiracy that block the progress of others because of race or creed, without regard to ability.

It is toward concern for these issues—and vigorous participation on the side of freedom—that our Irish heritage must impel us. If we are true to this heritage, we cannot stand aside.

There are two other areas of concern which I feel are of paramount importance and to which the Irish tradition speaks in ringing tones. One is the status of freedom in colonies, and second our relationship to the underdeveloped nations of the world.

The greatest enemy of freedom today, of course, is Communism, tyranny that holds its captives in viselike subjugation on a global scale. For nearly twenty years we and our allies have striven to halt the Communist advance. But one of the weaknesses in our common fear has been the restraint on freedom sponsored by our allies and accepted by ourselves.

The conduct of our foreign affairs should be consistently based on our recognition of every man's rights to be economically and politically free. This is in the American tradition. We were, after all, the victor in our own war for independence. We promulgated the Monroe Doctrine and the Open Door Policy with their clear warnings to the colonial powers of Europe.

We gave self-determination to our own dependencies, and for more than a century we opposed colonial exploitation elsewhere. But throughout all this we were still living largely in splendid isolation, removed from a direct control of world destiny. This was changed by World War II.

The frontiers of our national security became the

frontiers of the world. We found ourselves obliged to deal with the harsh facts of existence on a global basis. For the sake of our own security, we found our destiny to be closely linked with that of nations that maintained large colonial empires on which they felt their ultimate security depended. In some of the underdeveloped countries we have found our destiny linked with ruling powers or classes which hold the vast majority of their people in economic or military subjugation.

It is easy for us to believe that the imperialism of the West was infinitely preferable to the tyranny of Communism. But the sullen hostility of the African and Asian colonial nations has shown us that not all hold the same view. The bloody struggles for liberty from the sands of Algeria to the steaming jungles of Indonesia and Vietnam proved that others would make the same sacrifices to throw off the yoke of imperialism today that the Irish did more than half a century ago.

And we have a longer way to go in helping the people of some other nations to free themselves from economic domination. This is a part of our national policy not only because it is humane but also because it is essential. Our future may depend on how well this is understood throughout the world—how well it is understood that we still champion the quality of freedom everywhere that Americans enjoy at home.

I like to think—as did President Kennedy—that the emerald thread runs in the cloth we weave today, that these policies in which he believed so strongly, and which President Johnson is advancing, are the current flowering of the Irish tradition. They are directed toward freedom for *all* Americans here and for all peoples throughout the world. And I like to think that these policies will survive and continue as the cause of Irish freedom survived the death of the Liberator, Owen Roe O'Neill.

As you'll recall, O'Neill was one of the great figures of Irish history. It was of the period after his death, when the entire Irish nation was overwhelmed with grief, that the following lines were written:

Sagest in the council was he,
Kindest in the Hall;
Sure we never won a battle
'Twas Owen won them all.
Soft as woman's was your voice, O'Neill:
Bright was your eye,
Oh! why did you leave us, Owen?
Why did you die?

Your troubles are all over,
You're at rest with God on high,

But we're slaves, and we're orphans, Owen!
Why did you die?

We're sheep without a shepherd,
When the snow shuts out the sky—
Oh! why did you leave us, Owen?
Why did you die?

So on this St. Patrick's evening, let me urge you one final time to recall the heritage of the Irish. Let us hold out our hands to those who struggle for freedom today—at home and abroad—as Ireland struggled for a thousand years.

Let us not leave them to be "sheep without a shepherd when the snow shuts out the sky." Let us show them that we have not forgotten the constancy and the faith and the hope of the Irish.

Kennedy never would forget Scranton. On the plane back to Washington, astonished by the reception he had received—the poignant outpouring of love for President Kennedy and much evidence that he stood well with the people in his own right—he made an irrevocable decision about his future. Somehow, he would remain in public service.

California Institute of Technology

PASADENA, CALIFORNIA

JUNE 8, 1964

Immediately after President Kennedy's funeral, political pundits and politicians began speculating about Robert Kennedy's future: Would he be Johnson's running mate or seek to be governor of Massachusetts? In February 1964, a grass-roots Kennedy-for-vice-president movement gathered strength in New Hampshire as the first of the nation's presidential primary elections approached, and it caused such consternation in the White House that Kennedy had to discourage it publicly. Then came Scranton with its emotional demonstration of how people felt about the Kennedy brothers.

By April, a rising tide of political activity swept Kennedy into an unannounced, restrained bid for the vice presidential nomination, though he believed instinctively and, as it turned out, correctly, that relations between him and Johnson were so strained that he'd be the last person LBJ would choose unless the president was lagging in the polls.

Nevertheless, the tempo of his daily schedule returned to pre-assassination days. Beginning in Charleston, West Virginia, he toured poverty-stricken areas, spoke at colleges, and dedicated JFK memorials in all parts of the nation. He flew to West Berlin, accompanied by Ethel and his four oldest children, and spent four tumultuous days on an unofficial visit to Poland. Particularly notable was the commencement address he gave at the California Institute of Technology just before the trip to Europe.

Many years ago, Albert Einstein addressed the students of this institute. "It is not enough," he said, "that you should understand about applied science in order that your work may increase man's blessings." And he added, "Concern for man himself and his fate must always form the chief interest of all technical endeavors, concern for the great unsolved problems of the organization of labor and the distribution of goods—in order that the creations of our mind shall be a blessing and not a curse to mankind. Never forget this in the midst of your diagrams and equations."

Thus I come to talk to you, not as scientists, but as citizens, some of the best-educated, most rational, and most creative citizens of our country. Accordingly, you have a larger responsibility to apply the fruits of your education, your reason, and your creativity to the future of society, as well as the future of science.

There can be no greater concern for each of us, as citizens, than how wisely and how honorably our nation discharges its responsibilities of preserving peace and promoting freedom. And the first obligation, if we are to preserve an opening to the future, is to make sure that there will be a future at all.

The leaders of the world face no greater task than that of avoiding nuclear war. While preserving the cause of freedom, we must seek abolition of war through programs of general and complete disarmament. The Test Ban Treaty of 1963 represents a significant beginning in this immense undertaking.

We cannot pretend that such beginnings signal a millennium or armistice in the Cold War. They are modest steps. But they are steps forward, steps toward the ultimate goal of effective and reliable international controls over the destructive power of nations. Until such a goal can be achieved, however, we have no other choice than to insure that we can defend our country and help other peoples who are willing to work for their independence . . .

The United States must continue to expand its efforts to reach the peoples of other nations—particularly young people in the rapidly developing southern continents. Governments may come and go, but in the long run the future will be determined by the needs and aspirations of these young people . . .

Over the years, an understanding of what America really stands for is going to count far more than missiles, aircraft carriers, and supersonic bombers. The big changes of the future will result from this understanding—or lack of it.

We have made some progress in reaching the peoples of both countries . . . but the critical moves—the moves that will determine our success—are the kinds of political choices this country makes picking its friends abroad—and its enemies.

Far too often, for narrow tactical reasons, this country has associated itself with tyrannical and unpopular regimes that had no following and no future. Over the past twenty years, we have paid dearly because of support given to colonial rulers, cruel dictators, or ruling cliques void of social concern. This was one of President Kennedy's gravest concerns . . .

Ultimately, Communism must be defeated by progressive political programs which wipe out poverty, misery, and discontent on which it thrives. For that reason, progressive political programs are the best way to erode the Communist presence in Latin America, to turn back the Communist thrust in Southeast Asia, and to insure the stability of the new African nations and preserve stability in the world . . .

To say that the future will be different from the present is, to scientists, hopelessly self-evident. I observe regretfully that in politics, however, it can be heresy. It can be denounced as radicalism or branded as subversion. There are people in every time and every land who want to stop history in its tracks. They fear the future, mistrust the present, and invoke the security of a comfortable past which, in fact, never existed . . .

The danger of such views is not that they will take control of the American government. In time, the consensus of good sense which characterizes our political system will digest and discard frozen views and impossible programs. But there is a *short-term* danger from such voices . . .

The answer to these voices cannot simply be reason, for they speak irrationally. The answer cannot come merely from government, no matter how conscientious or judicious. The answer must come from within the American democracy. It must come from an informed, rational consensus which can recognize futile fervor and simple solutions for what they are—and reject them quickly.

A few days before the Cal Tech speech, Arizona Senator Barry Goldwater defeated New York Governor Nelson Rockefeller in the California Republican primary, clinching the GOP presidential nomination. At that point Kennedy knew Johnson would not need him, because Goldwater would not be a formidable threat to win in November. On July 29, after the Republican convention had nominated Goldwater, Johnson told Kennedy he would not be on the ticket. But by that time one of Kennedy's brothers-in-law, Stephen Smith, and some friends and associates were working to get a divided Democratic leadership in New York to close ranks and nominate Kennedy for United States senator.

Democratic National Convention

ATLANTIC CITY, NEW JERSEY

AUGUST 27, 1964

Kennedy formally entered the Senate race from the steps of Gracie Mansion, the official residence of New York City mayors, on Tuesday, August 25. Two days later, after the Democrats in convention at Atlantic City had nominated Johnson and Minnesota Senator Hubert Humphrey, Robert and Ethel Kennedy flew there to introduce a filmed memorial to JFK. That night, after a round of emotional meetings with state delegations to thank them for supporting his brother at the 1960 convention, Kennedy sat on the wooden steps behind the convention rostrum, nervously writing last-minute changes in his speech. What followed was unforgettable—a special moment in the history of American politics.

As Senator Henry Jackson of Washington introduced Kennedy and he came to the microphone, the delegates and audience stood. They applauded and cheered in wave after wave for twenty-two minutes. At first, Kennedy smiled wanly. He tried repeatedly to begin: "Mr. Chairman . . . Mr. Chairman . . ." but the cheering only intensified. Then he turned and looked helplessly at Jackson. Then he stood, eyes glistening, waiting for so much pent-up emotion to subside. When the hall became quiet, he started to speak, largely extemporaneously.

Mr. Speaker, Mr. Chairman, Mrs. Johnson, Senator Jackson, ladies and gentlemen, I wish to speak just for a few minutes.

I first want to thank all of you delegates to the Democratic National Convention and the supporters of the Democratic Party for all that you did for President John F. Kennedy. I want to express my appreciation to you for the efforts that you made on his behalf at the convention four years ago, the efforts that you made on behalf of his election in November of 1960, and, perhaps most importantly, the encouragement and the strength that you gave after he was elected president of the United States.

I know that it was a source of the greatest strength to him to know that there were thousands of people all over the United States who were together with him, dedicated to certain principles and to certain details.

No matter what talent an individual possesses, what energy he might have, no matter how much integrity and how much honesty he might have, if he is by himself, and particularly a political figure, he can accomplish very little. But if he is sustained, as President Kennedy was, by the Democratic Party all over the United States, dedicated to the same things that he was attempting to accomplish, he can accomplish a great deal.

No one knew that more than President John F. Kennedy. He used to take great pride in telling of the trip that Thomas Jefferson and James Madison made up the Hudson River in 1800 on a botanical expedition searching for butterflies; that they ended up down in New York City and that they formed the Democratic Party.

He took great pride in the fact that the Democratic Party was the oldest political party in the world, and he knew that this linkage of Madison and Jefferson with the leaders in New York combined the North and South and combined the industrial areas of the country with the rural farms. This combination was always dedicated to progress and all of our presidents have been dedicated to progress.

He thought of Thomas Jefferson and the Louisiana Purchase, and when Jefferson realized that the United States could not remain on the eastern seaboard he sent

Lewis and Clark to the West Coast; of Andrew Jackson; of Woodrow Wilson; of Franklin Roosevelt, who saved our citizens who were in great despair because of the financial crisis; of Harry Truman, who not only spoke but acted for freedom.

So when he became president, he not only had his own principles and ideals, but he had the strength of the Democratic Party. As president he wanted to do something for the mentally ill and the mentally retarded; for those who were not covered by Social Security; for those who were not receiving an adequate minimum wage; for those who did not have adequate housing; for our elderly people who had difficulty paying their medical bills; for our fellow citizens who are not white and who had difficulty living in this society. To all this he dedicated himself.

But he realized also that in order for us to make progress here at home we had to be strong overseas, that our military strength had to be strong. He said one time, "Only when open arms are sufficient, without doubt, can we be certain, without doubt, that they will never have to be employed." So when we had the crisis with the Soviet Union and the Communist Bloc in October of 1962, the Soviets withdrew their missiles and bombers from Cuba.

Even beyond that, his idea was that this country, that this world really, should be a better place when we turned it over to the next generation than when we inherited from the last generation. That is why—with all of the other efforts he made—the Nuclear Test Ban Treaty, which was done with Averell Harriman, was so important to him.

And that's why he made such an effort and was committed to the young people not only of the United States but to the young people of the world. And in all of these efforts you were there—all of you.

When there were difficulties, you sustained him.

When there were periods of crisis, you stood beside him. When there were periods of happiness, you laughed with him. And when there were periods of sorrow, you comforted him.

I realize that as individuals we can't just look back, that we must look forward. When I think of President Kennedy, I think of what Shakespeare said in *Romeo and Juliet:*

When he shall die
Take him and cut him out in little stars
And he will make the face of heaven so fine
That all the world will be in love with night,
And pay no worship to the garish sun.

I realize that as individuals and, even more important, as a political party and as a country, we can't just look to the past, we must look to the future.

So I join with you in realizing that what started four years ago, what everyone here started four years ago—that that's to be sustained; that that's to be continued. The same efforts and the same energy and the same dedication that was given to President John F. Kennedy must be given to President Lyndon Johnson and Hubert Humphrey.

If we make that evident, it will not only be for the benefit of the Democratic Party, but, far more important, it will be for the benefit of this whole country.

When we look at this film we must think what President Kennedy once said: "We have the capacity to make this the best generation in the history of mankind, or make it the last."

If we do our duty, if we meet our responsibilities and our obligations, not just as Democrats but as American citizens in our local cities and towns and farms and our states and in the country as a whole, then this generation of Americans is going to be the best generation in the history of mankind.

He often quoted from Robert Frost—and said it applied to himself—but we could apply it to the Democratic Party and to all of us as individuals:

The woods are lovely, dark and deep.
But I have promises to keep
And miles to go before I sleep,
And miles to go before I sleep.

PART THREE

The Senate Years

1965–1968

Introduction

Robert Kennedy received the cheers of the 1964 Democratic convention in his brother's stead, but the delegates' extraordinary outpouring of affection also reflected Robert's growing political appeal. The next four years would mark the busiest period in his life, and the sole time he would serve in elective office. He would give more than three hundred speeches, touching on virtually every topic of interest to him, including the seminal flashpoints of a decade accelerating in turbulence. And by 1968, Robert Kennedy would become America's best-known contemporary political figure, partly as a result of this intense public exposure.

Frequent appearances would help Kennedy refine his style, but his real power as a speaker during these years was derived principally from Kennedy's growing command of diverse policy matters and from his open, emotionally expressed sense of moral urgency in addressing the nation's central problems. John Kennedy's legacy would gradually assimilate into Robert's increasingly self-assured, independent public persona.

Robert Kennedy was also rising to the challenge of his times. It has been said of Winston Churchill that he inspired so much from his countrymen because he idealized *them* so completely; they stretched to meet his vision. Similarly, President Kennedy's assassination caused tremendous visceral grief for millions of Americans because in one senseless and unexpected moment, they lost not only their champion but the summoner of their own better angels. As the 1960s continued to unfold, the problems confronting Americans seemed increasingly complex and intractable. In Vietnam, neither the sacrifice of U.S. soldiers' lives nor the unleashing of awesome and unprecedented firepower brought victory. The world seemed composed of fickle allies and implacable foes, and dominating all was the certainty that superpower conflict could unleash global Armageddon.

At home, uneasy relations between races degenerated into mutual incomprehension and escalating violence. When the civil rights movement was confined to southern cities, most Americans experienced it through television images, and they were repulsed by the brutality they witnessed. But when both the venues for confrontation and the tactics of the combatants changed, the challenges became personalized—threatening attitudes, lifestyles, and personal safety.

Recognition of the country's challenges was widespread, but the extent of an individual's responsibilities seemed less clear, and whether *any* personal action could be effective seemed doubtful. Countrymen viewed one another across widening chasms of age, class, background, and beliefs. Because the stakes seemed elevated, tolerance of other views diminished.

It was a time, Kennedy told students at the University of Cape Town in 1966, "of danger and uncertainty; but [it was] also more open to the creative energy of men than any other time in history." And then he gave his listeners a challenge, which also provides the standard for evaluating Kennedy's own performance during his Senate years: "Everyone . . . will ultimately be judged—will ultimately judge himself—on the effort he has contributed to building a new world society and the extent to which his ideals and goals have shaped that effort."

No phase of Robert Kennedy's life provides such a wealth of public material. The evolution of Kennedy's views and the national (or international) context of his subject matter are clearly seen when his Senate speeches are organized thematically, rather than purely chronologically. Accordingly, seven sections are presented in this part: the 1964 New York Senate campaign; youth; blacks and the urban crisis; rural poverty and hunger; a selection of other domestic issues; diverse foreign policy concerns; and, finally, the overarching leitmotif of the second half of the decade—the war in Vietnam.

Robert Kennedy on His Own

The 1964 Senate Race

Ever since his brother's assassination, the press had considered it inevitable that Robert Kennedy would be a candidate for *some* office, as the prelude to an equally certain run for the presidency, but Kennedy was hardly a natural candidate. His strategic and administrative skills were unquestionable after his management of John Kennedy's presidential campaign, but Robert Kennedy seemed temperamentally best suited to such behind-the-scenes involvement. As his brother's manager, his brusqueness alienated powerful party leaders he felt weren't productive; on the other hand, his inherent shyness made him an occasionally uncomfortable public speaker and a sometimes clumsy campaigner.

But in a nation still stunned and grieving over the death of John Kennedy, brother Robert had enormous appeal. He was enough of a tactician to sketch the contours of the upcoming campaign accurately. He told Ed Guthman (who resigned from the Justice Department when Kennedy did, to help guide the New York effort), "I'll draw huge crowds as I go to different parts of the state for the first time. All the attention will be on that, and it will last for about three weeks. I'll hit a low point around the first of October. The question will be whether I can turn it around and regain the momentum."

Kennedy's prediction was accurate. Certain elements of risk appeared early: Kennedy was branded a carpetbagger; the liberal community felt uneasy, especially with his youthful service on the staff of the Senate Permanent Investigations Subcommittee headed by Senator Joe McCarthy; the regular party organization was riven with factional strife between the traditional and reform wings; and the state's most influential paper, the *New York Times,* was hostile.

Nothing, however, dampened the enthusiasm of the crowds swarming around the candidate at every stop. On his initial campaign swing upstate (the home turf of his incumbent opponent, Kenneth Keating), 150,000 New Yorkers of all ages waited two hours along Kennedy's route into Buffalo to greet him when he tardily arrived. He was even later for his last stop the next day, in Glens Falls (population 21,000), and Kennedy was certain no one would be waiting. But one thousand people, many with children in their arms, cheered the arriving candidate at the airport, and a band struck up a welcoming tune. Kennedy thanked the crowd, grinned and said, "Well, here we are, five hours late. That's the smooth, hard-driving, well-oiled Kennedy machine for you." The road into town was lined with parents and children, many in their bedclothes, waving and shouting greetings. Four thousand people waited in the town square. It was one thirty in the morning.

With the election less than two months away, Keating counterattacked, claiming that "people have been standing in line at the World's Fair [held in New York City that year] longer than my opponent has been a resident of New York." He argued that Kennedy would brush aside six years of Keating's service to the state in his "ruthless" quest for power.

And the incumbent moved to undercut Kennedy's political base by harshly attacking the latter's character: before black audiences, charging Kennedy with running out on the civil rights struggle by leaving the Justice Department; to Italian groups, arguing that Kennedy's prosecution of Mafia figures slurred Italian

Columbia/Barnard Democratic Club

NEW YORK, NEW YORK

OCTOBER 5, 1964

Nevertheless, Keating's relentless attacks succeeded in making the race nearly even by early October, when Kennedy fielded questions from students and others at an evening event at Columbia University, in New York City. Kennedy's responses to questions reflecting the principal points of controversy in the Senate race foreshadowed many of the major areas of conflict that would demand Kennedy's attention in the coming years.

Q.—Sir, could you tell us why you chose to run from New York rather than from Massachusetts or Virginia or someplace else? [*Applause.*]

A.— . . . If the election's going to be decided based on my accent or where I also have other associations, I think that then people are going to vote for my opponent. I think that the election really should be decided on the basis of [whether] . . . my opponent or myself can do more for the state of New York and make the greatest amount of difference for this state and for the country over the period of the next six years.

I have been involved in matters which intimately affect the state of New York, whether it's the question of war or peace, whether we're going to survive or not, the problems of Southeast Asia, or the problems of Berlin, or nuclear control, the establishment of the Peace Corps—all of these matters in which I've been intimately involved . . .

Second, on the question of domestic matters, the problem of what we're going to do about our young people, what we're going to do about education, what we're going to do about unemployment, what we're going to do about housing—all of these matters, again, are questions about which I have strong opinions and in which I've been involved, over the period of the last three and a half years, and in which we made an effort to do something about.

I don't say that we made these problems disappear. But I was involved in them, and we accomplished something, whether it was the biggest housing bill that was passed in the history of the United States, in 1961; whether it was the fact that we passed more legislation in the field of education than at any time since Abraham Lincoln was president . . . or whether it was the civil rights bill. We passed more crime legislation than has ever been passed. We passed—for the first time, they've been talking about providing a lawyer for the indigent and the poor, and nobody ever did anything about it—we did something about it. I think that's what a United States senator should do. I think it's more than a question of just showing up for roll call. I think it's a question of getting some leadership, getting some direction, getting some things done . . .

Again, if it's going to be judged on who's lived here in the state of New York longer, then my opponent has. But then maybe you should elect the oldest man in the state of New York. [*Applause.*] . . .

I have [had] really two choices over the period of the last ten months. I could have stayed in—I could have retired. [*Laughter.*] And I—my father has done very well and I could have lived off him. [*Laughter and applause.*] Or I could have continued to work for the government. I mean that's my major interest and it's been the major interest of my family . . .

I tell you frankly I don't need this title because I [could] be called general, I understand, for the rest of my life . . . [*Laughter and applause.*] And I don't need the money and I don't need the office space . . . [*Laughter.*] . . . Frank as it is—and maybe it's difficult to understand in the state of New York—I'd like to just be a good United States senator. I'd like to serve. [*Applause.*] . . . Let me just say: I run for the United States Senate because I think it's such an important position;

because I want to play a role. I saw over the period of the last three and a half years . . . what a difference an individual can make . . .

President Kennedy said: "We have the capacity to make this the best generation in the history of mankind or make it the last." I think that we can make it the best generation. I think, really, it's going to rest with those who are educated, those people who are trained—whether they're going to participate, or whether they're going to say this is the problem or responsibility of somebody else. That's why I think that all of you who have had the advantage of an education here, just as I've had the advantage of an education, that we have a special and particular responsibility. When we look and analyze where our government came from . . . we think back to the Greeks, what their idea really was of participation, and what Pericles said in his funeral oration: "We differ from other states in that we regard the individual who holds himself aloof from public affairs as being useless. Yet we yield to no one in our independence of spirit and complete self-reliance."

I think that's what's best to guide us. The Greek word *idiot* comes from that individual who didn't participate, who wasn't actively involved. President Kennedy's favorite quote was really from Dante: "The hottest places in hell are reserved for those who in time of moral crisis preserve their neutrality."

So it's all of us—whether it's in the field of civil rights or housing or whether it's in Vietnam or whatever it is . . . If we're going to permit what's going on in Harlem now, of those young children who grow up uneducated and untrained and dissatisfied with life and dissatisfied with their future and feeling that there is nothing in this system, then we're going to be in difficulty. Even if we look at this selfishly, we're going to be in difficulty . . .

Sophocles said one time: "Well, what joy is there in day that follows day, some swift, some slow, with death the only goal?" Really, that's what many of our fellow citizens feel—whether they're in Appalachia or whether they're in Harlem or whether they're the white children who live in some of these other areas where they've no future. Or even our elderly people . . . who feel that there's not any future for them, that all they're going to do is die and get sick and go have to take a pauper's oath. These, I think, are our responsibility. These are your responsibilities, just as they are mine.

Now, I hope that I win as United States senator, but even if I don't, I think that for all of us, that we have an obligation, we have a responsibility. If we don't do it, then nobody is going to do it . . . And we have a special—not only responsibility—but a special opportunity to make a difference in the world, and make a difference for this country. And that's really what I've come to say to you. I think of what Archimedes said, explaining the lever: "Show me where I can stand and I can move the world." And I think that we can. I think that we started three and a half years ago and I think that we can continue it. And I don't think it's just a question of political belief. I think that we can make a difference.

Kennedy began to regain the campaign's momentum, allying himself with the popular Johnson-Humphrey ticket, strengthening his campaign organization, and attacking Keating's record. Three additional factors were critical to Kennedy's success: effective use of television advertising, including a half-hour segment culled from the Columbia event; the loyalty of black voters, despite some black leaders' support of Keating; and an impressive performance in the only candidates' "debate." The latter was a fiasco: a debate had been long delayed as the candidates jockeyed for position; then, a week before the election, Keating decided to debate an empty chair on television but panicked in full view of the media when Kennedy appeared at the studio door and was refused entry. Humiliated, Keating rejected any debate, and his polling numbers plummeted.

In the end, Kennedy won by more than seven hundred thousand votes, considerably less than Johnson's margin in the presidential race, but nearly twice the margin the new senator had expected. For only the second time in history, two brothers would serve together in the United States Senate: Younger brother Edward M. Kennedy had been elected from Massachusetts two years earlier.

Robert Kennedy and the Young

It wasn't unusual that Kennedy made the Columbia session a focal point for his ads, for he enjoyed student audiences even before becoming attorney general. While in his brother's cabinet, he had used colleges and universities to address major public policy topics, as at the University of Georgia Law School in 1961, where he presented the administration's civil rights policy to the Deep South. He was an even more frequent visitor to campuses as a senator.

Part of the Kennedys' electoral appeal was their vigorous image and relative youth: John Kennedy was inaugurated at forty-three, and Robert ran for the Senate at thirty-eight. They were among the youngest national leaders since the founding of the republic. Throughout President Kennedy's administration, no member of this most photographed of families seemed ever to be pictured at rest but always playing football, sailing, or emerging from an ocean swim. Robert Kennedy was the most competitive of the lot.

President Kennedy, having read Theodore Roosevelt's 1908 proposition that U.S. Marines should be able to hike fifty miles in twenty hours, challenged the modern Marines to the test. Robert Kennedy, accompanied by Ed Guthman (a combat veteran) and three others, embarked totally unprepared on just such a marathon hike in 26-degree weather in February 1963. Kennedy made it—but one by one his companions were forced to drop off.

Kennedy felt a kinship with the baby boom generation moving through the nation's colleges by the mid-1960s. When Kennedy entered the Senate, his two legislative assistants, Peter Edelman and Adam Walinsky, were twenty-seven and twenty-eight, respectively, and much of the political and personal creative tension that marked the remainder of his career arose because Robert Kennedy increasingly functioned as a bridge between the young and the World War II veterans of his generation.

At their best, students encouraged Kennedy with their idealism and inquisitiveness. He, in turn, habitually challenged them, and when he confronted campus audiences seemingly frozen in their material comfort, the exchanges became particularly sharp and moving.

Three speeches during his Senate years demonstrate Kennedy's format before student audiences: an exposition of a critical policy issue, followed by a call to personal involvement. Presented in this section are addresses at Queens College in New York City (June 15, 1965), which opened with a description of the revolutionary spirit sweeping the Third World; at the University of Mississippi (March 18, 1966), where a southern audience was urged to rise to the opportunity posed by civil rights; and at the University of California at Berkeley (October 22, 1966), the birthplace of the Free Speech Movement, where Kennedy called for disciplining dissent in the service of racial justice.

Finally, speaking to a middle-age audience at an Americans for Democratic Action dinner in February 1967, Kennedy carried the banner of the young in a discussion of the growing estrangement between generations.

Commencement Address at Queens College

NEW YORK, NEW YORK

JUNE 15, 1965

Conscious of the carpetbagger charges that dogged him during the 1964 campaign, Kennedy spent most of his first year in the Senate either in Washington or in New York. Commencement at Queens College in New York City provided an opportunity to address themes he returned to a number of times in 1965: the wave of change sweeping the globe—from the rising expectations in the developing countries to the struggle for civil rights in America—forces that demanded increased public participation (particularly by the educated) in the issues of the times.

Around the world—from the Straits of Magellan to the Straits of Malacca, from the Nile delta to the Amazon basin, in Jaipur and Johannesburg—the dispossessed people of the world are demanding their place in the sun. For uncounted centuries, they have lived with hardship, with hunger and disease and fear. For the last four centuries, they have lived under the political, economic, and military domination of the

West. We have shown them that a better life is possible. We have not done enough to make it a reality.

A revolution is now in progress. It is a revolution for individual dignity, in societies where the individual has been submerged in a desperate mass. It is a revolution for self-sufficiency, in societies which have been forced to rely on more fortunate nations for their manufactured goods and their education, cotton textiles, and calculus texts. It is a revolution to bring hope to their children, in societies where 40 percent of all children die before reaching the age of five. This revolution is directed against us—against the one third of the world that diets while others starve; against a nation that buys eight million new cars a year while most of the world goes without shoes; against developed nations which spend over one hundred billion dollars on armaments while the poor countries cannot obtain the ten to fifteen billion dollars of investment capital they need just to keep pace with their expanding populations . . .

It is a revolution not just for economic well-being but for social reform and political freedom, for internal justice and international independence.

We should understand the legitimacy of these ideals—for they are only what our own forefathers sought. We should recognize their power—for they have sustained us throughout our history . . .

We must remember our revolutionary heritage. We must dare to remember what President Kennedy said we could not dare to forget—that we are the heirs of a revolution that lit the imagination of all those who seek a better life for themselves and their children; that we must seize the chance to lead this continuing revolution, not block its path; that we must stand, not for the status quo, but for progress; that we must practice abroad what we preach at home . . .

The essence of the American Revolution—the principle on which this country was founded—is that direct participation in political activity is what makes a free society.

Freedom, for the founders, was not merely negative, the absence of arbitrary restraints. Freedom for them was active and positive—the power of each individual to take part in the government of the town, the state, the nation. As Jefferson said, "not merely at an election one day in the year, but every day," every man was to be "a participator in the government of affairs." As time has passed, and society has grown more complex, this tradition has been difficult to maintain . . .

But the sit-ins and the teach-ins, the summer projects, the civil rights vigils and civil liberties protests, organizing the poor and marching on Washington—all these may be helping to return us to a politics of public participation, where individual citizens, without holding political office, may still contribute to the public dialogue—where they do something more than write letters to the newspaper or answer *yes* or *no* on a public opinion poll . . .

Your activity, moreover, is to be welcomed whether any of us think your opinions are right or not—for we develop truth and policy in the midst of debate.

But if you are to participate—if you are to help lead this nation—you must use, to the limits of your power, the education you have been given here. This education has taught you the value of facts—and there can be no meaningful politics which is ignorant of the facts. Your education has taught you that undue self-interest is the enemy of truth—and politics is corrupted where selfishness takes precedence over respect for the truth . . .

Second, you must consider and judge your acts with considerable care, and with a sense of responsibility. For your participation . . . imposes special obligations.

The Berkeley students of the Free Speech Movement made a contribution to academic freedom and help also to remind universities all over the country that schools are for teaching.

But when a few students turned Free Speech into scrawling of dirty words on placards, they discredited not only themselves but the initial protest . . .

It is not helpful—it is not honest—to protest the war in Vietnam as if it were a simple and easy question, as if any moral man could reach only one conclusion. Vietnam admits no simple solution. But the complexity and difficulty of any question should not keep you from speech or action . . .

In 1787, Benjamin Rush said, "There is nothing more common than to confound the term of *American Revolution* with that of the late *American War.* The American War is over; but this is far from being the case with the American Revolution. On the contrary, nothing but the first act of the great drama is closed."

That is as true today as it was then. For as long as men are hungry, and their children uneducated, and their crops destroyed by pestilence, the American Revolution will have a part to play. As long as men are not free—in their lives and their opinions, their speech and their knowledge—that long will the American Revolution not be finished . . .

University of Mississippi Law School Forum

OXFORD, MISSISSIPPI

MARCH 18, 1966

It was easy, physically and politically, for Robert Kennedy to choose to address Queens College students in his adopted state of New York; appearing at the University of Mississippi nine months later was entirely different. Kennedy's consideration of the speaking invitation there, the tone and substance of the remarks he prepared, and the reception he received were critically affected by a pivotal incident that occurred while he was attorney general. Less than four years earlier, the Ole Miss riots around James Meredith's enrollment were the worst federal-state strife since the Civil War, and many in the Deep South blamed the then–U.S. attorney general—Robert Kennedy—and his brother, who had sent in federal troops.

When law students at the University of Mississippi arranged for Robert Kennedy to be invited to speak on campus in 1966, the senator and his staff asked for advice from friends throughout the South. Senate aide Peter Edelman remembered that "the senator would say 'What do they say down there? . . . Is there anything I can say? . . . Will there be any common ground at all?'" Finally, in late February 1966, Kennedy decided to go.

Immediately, the FBI reported threats against his life, but Kennedy pressed on, asking Edelman and Walinsky to work with two of his brother's speech writers (historian Arthur Schlesinger Jr. and Dick Goodwin) on a draft. Kennedy reviewed and revised the collaborative effort, still uncertain of the reception he'd receive.

Kennedy was accompanied by his friend the writer Peter Maas, who recalled that the two flew in a small plane from Memphis, "bucking up and down over scraggly Mississippi pine growths as [they] skirted a line squall." The scene on arrival stunned them: "Literally hundreds of students," Maas remembered, "raced forward . . . and cheered him wildly. While this was going on, [Maas] happened to be standing next to a goggle-eyed Mississippi state trooper, the perfect caricature of the . . . southern cop, booted and big-bellied, feet apart, thumbs hooked into his belt. The trooper said it all. 'Ah don't believe it,' he muttered to himself, 'but Ah'm seein' it.'"

Your generation—South and North, white and black—is the first with the chance not only to remedy the mistakes which all of us have made in the past but to transcend them. Your generation—this generation—cannot afford to waste its substance and its hope in the struggles of the past, for beyond these walls is a world to be helped, and improved, and made safe for the welfare of mankind . . .

Here in America . . . we must create a society in which Negroes will be as free as other Americans—free to vote, and to earn their way, and to share in the decisions of government which shape their lives. We know that to accomplish this end will mean great tension and difficulty and strife for all of us, in the North as much as in the South. But we know we must make progress, not because it is economically advantageous, not because the law says to, but because it is right.

Change is crowding our people into cities scarred by slums—encircled by suburbs which sprawl recklessly across the countryside; where movement is difficult, beauty rare, life itself more impersonal, and security imperiled by the lawless.

And beyond this, modern change is assaulting the deepest values of our civilization—those worlds within a world where each can find meaning and importance and warmth: family and neighborhood, community and the dignity of work.

Family ties grow weaker as the span between the generations widens. The community, a haven where each could once find warmth and significance, begins to dissolve as the streets of the cities rush in upon each other. Work becomes mechanical and routine, eroding the self-respect which individual effort once provided. And, especially here in the South, rapidly shifting relations between the races destroy old certainties and demand new attitudes and values.

And if our nation is changing, the world around us is moving even faster. Since I graduated from high school, the United States has fought in two wars and is now involved in a third. Nuclear weapons have been invented and tested, and they have now spread to five nations. The great colonial empires of Europe have been dissolved, and more than seventy new nations have now been created. We begin to learn how to deal with one great hostile power, the Soviet Union; and then beyond its borders, the empire of China begins to grow in significance and in danger.

And in every continent—from Jaipur to Johannesburg, from Point Barrow to Cape Horn—men claim their right to share in the bounty which modern knowledge can bring, and they claim also the justice which they have heard proclaimed in that document which listed the inalienable rights of man. I have seen scrawled on the sidewalks of Indonesia, and on the walls of Africa and in Latin America, not WORKERS OF THE WORLD UNITE, but ALL MEN ARE CREATED EQUAL, and GIVE ME LIBERTY OR GIVE ME DEATH. They draw their hope for change and for a better life from the example of the United States. They look to us for hope and for help. And the real question before you, before all young Americans, is whether we will help bring about that future, or whether we will not help and stand by.

In such a challenging world—such a fantastic and dangerous world—we will not find answers in old dogmas, by repeating outworn slogans, or fighting on ancient battlegrounds against fading enemies long after the real struggle has moved on. We ourselves must change in order to master change. We must rethink our old ideas and beliefs before they capture and before they destroy us. For those answers America must look to its young people, the children of this time of change. And we look especially to that privileged minority of educated men and women who are the students of this country.

For the answers we seek must be found in the light of reason—by fact and logic and careful thought, unsustained by violent prejudices or by myths. And those are the answers which your education has equipped you—more than any other group in this country—has equipped you to find . . .

Plato said that if we are to have any hope for the future, that those who have lanterns will pass them on to others. You must use your lamps, the lamps of your learning, to show our people past the forest of stereotypes and slogans into the clear light of reality and of fact and of truth . . .

It is not enough to understand, or to see clearly. The future will be shaped in the arena of human activity, by those willing to commit their minds and their bodies to the task. Ralph Waldo Emerson said, "God gives to each of us the choice between truth and repose. Take which you please. You cannot have both." . . .

You will go as ambassadors of the twentieth century to a past struggling toward the possibilities of the modern world. And in so doing you serve not only man but the cause of national growth and of national independence, which is the foundation of a peaceful world order . . .

Oliver Wendell Holmes once said, "As life is action and passion, it is required of a man that he should share the passion and action of his time, at peril of being judged not to have lived."

It is simple to follow the easy and familiar path of personal ambition and private gain. It is more comfortable to sit content in the easy approval of friends and of neighbors than to risk the friction and the controversy that comes with public affairs. It is easier to fall in step with the slogans of others than to march to the beat of the internal drummer—to make and stand on judgments of your own. And it is far easier to accept and to stand on the past than to fight for the answers of the future.

But . . . Goethe told us that Faust lost the liberty of his soul when he said to the passing moment, "Stand, thou art so fair." And there would be no surer way for us to lose our liberty and the true meaning of our heritage than to make the same mistake.

And each of us will ultimately be judged—and will ultimately judge himself—on the extent to which he personally contributed to the life of this nation and to world society of the kind we are trying to build.

Jefferson Davis once came to Boston and he addressed his audience in Faneuil Hall as "countrymen, brethren, Democrats." Rivers of blood and years of darkness divide that day from this. But those words echo down to this hall, bringing the lesson that only as countrymen and as brothers can we hope to master and subdue to the service of mankind the enormous forces which rage across the world in which all of us live. And only in this way can we pursue our personal talents to the limits of our possibility—not as northerners or southerners, black or white—but as men and women in the service of the American dream.

Kennedy received frequent and sustained applause from the audience of more than five thousand. Perhaps more surprising, the reaction was equally warm when, in response to a question after the address, he related in detail his actions and motivations in the Meredith registration battle three and a half years earlier. His deadpan account of Governor Barnett's proposals for intricate, staged confrontations between Barnett and federal agents was greeted, to Kennedy's astonishment, with peals of laughter from the Ole Miss students. Oscar Carr, a liberal Delta planter, recalled that Kennedy came to the Oxford campus "persona non grata . . . [yet] I have never seen a politician of any ilk, stature, or office, at any time in the state, more wildly acclaimed."

Speech to Students

UNIVERSITY OF CALIFORNIA AT BERKELEY

OCTOBER 22, 1966

As the midterm congressional elections loomed in 1966, Kennedy found himself in great demand from Democratic candidates throughout the country. In fact, Kennedy had moved dramatically in front of President Johnson in opinion polls questioning Democrats and Independents on their 1968 presidential preference. On August 21, 1966, pollster George Gallup commented on his just-concluded national survey that Kennedy's "rise in political appeal . . . has been spectacular."

In October, Kennedy embarked on a prolonged barnstorm for congressional candidates in the West. Included in the schedule was the Kennedy staple: a policy address delivered on a college campus, this time at the University of California at Berkeley.

Berkeley had been a hotbed of dissent, particularly since the 1964 Free Speech Movement, which sparked the first mass sit-ins on an American campus. That standoff lasted most of the fall academic term, after Chancellor Edward Strong denied the local chapter of the Congress of Racial Equality a venue on campus to recruit for protest activities. Widespread civil disobedience culminated in the occupation of the school administration building and the arrest of more than seven hundred demonstrators.

In 1966, shortly before Kennedy came to campus, former Supreme Court Justice and United States Ambassador to the United Nations Arthur J. Goldberg was forced to cut short his speech at Berkeley by hecklers, many carrying signs branding Goldberg a "Doctor of War" for his participation in the Johnson administration's policy in Vietnam.

Kennedy appeared at Berkeley after a full day of campaigning, including stops in the San Joaquin Valley and San Jose. Berkeley's outdoor Greek Theatre was jammed as he rose to the lectern.

The future does not belong to those who are content with today, apathetic toward common problems and their fellow man alike, timid and fearful in the face of new ideas and bold projects. Rather it will belong to those who can blend passion, reason, and courage in a personal commitment to the ideals and great enterprises of American society. It will belong to those who see that wisdom can only emerge from the clash of contending views, the passionate expression of deep and hostile beliefs. Plato said: "A life without criticism is not worth living."

This is the seminal spirit of American democracy. It is this spirit which can be found among many of you. It is this which is the hope of our nation.

For it is not enough to allow dissent. We must demand it. For there is much to dissent from.

We dissent from the fact that millions are trapped in poverty while the nation grows rich.

We dissent from the conditions and hatreds which deny a full life to our fellow citizens because of the color of their skin.

We dissent from the monstrous absurdity of a world where nations stand poised to destroy one another, and men must kill their fellow men.

We dissent from the sight of most of mankind living in poverty, stricken by disease, threatened by hunger, and doomed to an early death after a life of unremitting labor.

We dissent from cities which blunt our senses and turn the ordinary acts of daily life into a painful struggle.

We dissent from the willful, heedless destruction of natural pleasure and beauty.

We dissent from all those structures—of technology and of society itself—which strip from the individual the dignity and warmth of sharing in the common tasks of his community and his country.

These are among the objects of our dissent. Yet we

must, as thinking men, distinguish between the right of dissent and the way we choose to exercise that right. It is not enough to justify or explain our actions by the fact that they are legal or constitutionally protected. The Constitution protects wisdom and ignorance, compassion and selfishness alike. But that dissent which consists simply of sporadic and dramatic acts sustained by neither continuing labor or research—that dissent which seeks to demolish while lacking both the desire and direction for rebuilding, that dissent which contemptuously or out of laziness casts aside the practical weapons and instruments of change and progress—that kind of dissent is merely self-indulgence. It is satisfying, perhaps, to those who make it.

But it will not solve the problems of our society. It will not assist those seriously engaged in the difficult and frustrating work of the nation. And when it is all over, it will not have brightened or enriched the life of a single portion of humanity in a single part of the globe.

All of us have the right to dissipate our energies and talent as we desire. But those who are serious about the future have the obligation to direct those energies and talents toward concrete objectives consistent with the ideals they profess. From those of you who take that course will come the fresh ideas and leadership which are the compelling needs of America.

Devoted and intelligent men have worked for generations to improve the well-being of the American people, diminish poverty and injustice, and protect freedom. Yet even as we honor their accomplishments we know that our own problems will not yield to the ideas and programs on which past achievement has been built. Ideas are often more confining, more difficult to discard, in their success than in their failure. Yet we must now cast aside many tested concepts in the face of challenges whose nature and dimension are more complex and towering than any before. For this we must look to your generation, a generation which feels most intensely the agony and bewilderment of the modern age, and which is not bound to old ways of thought . . .

The great challenge before us is what you have gathered to consider: the revolution within our gates; the struggle of Negro Americans for full equality and freedom.

That revolution has now entered a new stage, one that is at once more hopeful and more difficult, more important and more painful. It is to give every Negro the same opportunity as every white man to educate his children, provide for his family, live in a decent home, and win human acceptance as well as economic achievement in the society of his fellows. And it is to do all this in the face of the ominous growth of renewed hostility among the races.

This will not be achieved by a law or a lawsuit, by a single program or in a single year. It means overcoming the scarred heritage of centuries of oppression, poor education, and the many obstacles to fruitful employment. It means dissolving ghettos—the physical ghettos of our big cities and those ghettos of the mind which separate white from black with hatred and ignorance, fear and mistrust . . .

Some among us say the Negro has made great progress—which is true; and that he should be satisfied and patient—which is neither true nor realistic. In the past twenty years we have witnessed a revolution of rising expectations in almost every continent. That revolution has spread to the Negro nation confined within our own. Men without hope, resigned to despair and oppression, do not make revolutions. It is when expectation replaces submission, when despair is touched with the awareness of possibility, that the forces of human desire and the passion for justice are unloosed . . .

We have held out the promise that color shall no longer stand in the way of achievement or personal fulfillment or keep a man from sharing in the affairs of the country. We have unveiled the prospect of full participation in American society, while television, radio, and newspapers bring to every Negro home the knowledge of how rewarding such participation can be. With so bountiful a promise, how much greater must be the frustration and the fury of the Negro—especially the young Negro . . . For him the progress of the past can count for little against the crushing awareness that his hopes for the future are beyond his reach for reasons which have little to do with justice or his worth as a man. Occasionally, broken hope and a deeply felt futility erupt in violence and extreme statements and doctrines. If we deny a man his place in the larger community then he may turn inward to find his manhood and identity, rejecting those he feels have rejected him . . .

But if any man claims the Negro should be content or satisfied, let him say he would willingly change the color of his skin and go to live in the Negro section of a large city. Then, and only then, has he a right to such a claim.

Yet however much the condition of most Negroes must call forth compassion, the violence of a few demand condemnation and action . . .

To understand the causes is not to permit the result. No man has the right to wantonly menace the safety and well-being of his neighbors. All citizens have the right to security in the streets of their community—in

Birmingham or in Los Angeles. And it is the duty of all public officials to keep the public peace and bring to justice those who violate it.

I know many of you understand the terrible frustration, the feeling of hopelessness, the passion for betterment which, denied to others, has turned to violence and hate. It is difficult to live in the shadow of a multimillion-dollar freeway, to watch the white faces blur as they speed by the problems of the city, returning each evening to the pleasant green lawns of the suburbs. And it must be difficult beyond measure to share in America's affluence enough to own a television set—and to see on that set the hate and fear and ugliness of little Negro children being beaten and clubbed by hoodlums and thugs in Mississippi.

Some have turned to violence. And the question many Negroes surely ask themselves—the question many of you surely ask yourselves—is, Why not?

Why not turn to violence? . . .

But the course of violence would be terribly, awfully wrong: Not just because hatred and violence are self-defeating—though they are self-defeating, for they strike at the very heart of obedience to law, peaceful process, and political cooperation which are man's last best hopes for a decent world.

We must oppose violence not because of what violence does to the possibilities of cooperation between whites and blacks; not just because it hampers the passage of civil rights bills, or poverty legislation, or open-occupancy laws.

The central disease of violence is what it does to all of us—to those who engage in it as much as to those who are its victims.

Cruelty and wanton violence may temporarily relieve a feeling of frustration, a sense of impotence. But the damage of those who perpetrate it—these are the negation of reason and the antithesis of humanity, and they are the besetting sins of the twentieth century.

Surely the world has enough, in the last forty years, of violence and hatred. Surely we have seen enough of the attempt to justify present injustice by past slights, or to punish the unjust by making the world more unjust.

We know now that the color of an executioner's robe matters little. And we know in our hearts, even through times of passion and discontent, that to add to the quantity of violence in this country is to burden our own lives and mortgage our children's souls, and the best possibilities of the American future.

If this is a challenge to the Negro community, and especially to the political courage of Negro leadership whose own position may be endangered by rising militance, the challenge to white America is equally great.

In recent months we have seen comment on what some have called the *backlash.* Opposition to violence and riots and irresponsible action is the justified feeling of most Americans, white and black. But that backlash which masks hostility to the swift and complete fulfillment of equal opportunity and treatment, which contains opposition to demands for justice and freedom, which denies the need to destroy slums, provide education, and eliminate poverty—that is wrong, shameful, immoral, and self-defeating. Any leader who seeks to exploit this feeling for the momentary advantage of office fails his duty to the people of this country.

It would be a national disaster to permit resentment, or fear at the actions of a few, to drive increasing numbers of white and black Americans into opposing camps of distrust and enmity . . . Some say that in the last analysis, after all, we need not fear injustice; that if our great common purpose divides into conflict and contest, the whites will win. In one sense, that is true. We are far more numerous and more powerful. But it would be a Pyrrhic victory. The cost would be decades of agony and civil strife, the sacrifice of our ideal of liberty, and ultimately the loss of the soul of our nation.

We can understand the apprehension of those white Americans who feel threatened in their persons or their property. Yet they are only being asked to permit others what they demand for themselves: an equal chance to share in the American life. The whole experience of our nation shows that as each minority emerged, those who came before feared damage to their own way of life, and that each time they were wrong. The achievements of each group enlarged the prospects of all. In President Kennedy's words: "The rising tide lifts all the boats." That will be our experience with the Negro too.

Moreover, we must all understand that the problem will not go away . . . Thus we have only one choice. We can face our difficulties and strive to overcome them; or we can turn away, bringing repression, steadily increasing human pain and civil strife, and leaving a problem of far more terrifying and grievous dimensions to our children. Anyone who promises another course, who pledges a solution without cost, effort, or difficulty, is deluding both himself and the people to whom he speaks.

Like other minority groups, Negroes will bear the major burden of their own progress. They will have to make their own way, as they are doing. But we must remember that other minorities, including my own, also made progress through increasing their political and economic power as well as by individual effort. Nor was that progress completely without violence, fear, and hatred. Moreover, earlier immigrants often began their cities by moving to the unsettled West, a door now closed; or finding unskilled labor, a door which is swiftly narrowing. Today, to find a job requires increasingly complex skills, denied to those without education. Nor did other minorities suffer under the special handicaps of the Negro heritage and the crushing forces of racial feeling from whose poisons few whites have fully liberated themselves.

Thus the changed circumstances of modern life and the peculiar nature of the Negro experience make large-scale government action necessary if we are to crush the remaining barriers to equal opportunity and to lead an accelerating national effort to give Negroes a fair chance to share equally in the abundance and dignity of American life . . .

Even if we do all this and much more, if we act on an unprecedented scale, progress will still be slow. It is true, as Jefferson wrote, that "the generation which commences a revolution rarely completes it." The problem of giving content to equality is deeply embedded in the structure of American life. It cannot be swept away with a single blow. Yet we can create the steady, concrete, and visible achievements which will justify and sustain the expectation that each year will bring greater opportunity than the last. And we can support and nourish the faith of Negro Americans that their country recognizes the justice of their cause and the urgency of their needs.

This is one of the many crossroads at which American life now stands. In the world and at home, you have the opportunity and the responsibility to help make the choices which will determine the greatness of this nation. You are a generation which is coming of age at one of the rarest moments in history—a time when all around us the old order of things is crumbling and a new world society is painfully struggling to take shape. If you shrink from this struggle, and these many difficulties, you will betray the trust which your own position forces upon you.

You live in the most privileged nation on earth. You are the most privileged citizens of that privileged nation, for you have been given the opportunity to study and learn, to take your place among the tiny minority of the world's educated men. By coming to this school you have been lifted onto a tiny, sunlit island while all around you lies an ocean of human misery, injustice, violence, and fear. You can use your enormous privilege and opportunity to seek purely private pleasure and gain. But history will judge you, and, as the years pass, you will ultimately judge yourself, on the extent to which you have used your gifts to lighten and enrich the lives of your fellow man. In your hands, not with presidents or leaders, is the future of your world and the fulfillment of the best qualities of your own spirit.

At the conclusion of his speech, as was his custom, Kennedy fielded questions, and one of the first was about the war in Vietnam. Kennedy had said little publicly about the conduct of the war since the early months of 1966, largely because the press interpreted every criticism as part of a calculated challenge to unseat Johnson.

Kennedy's press secretary, Frank Mankiewicz, who accompanied his boss on the California tour, remembered that questions about the war were asked at almost every stop, and Kennedy "was always very soft on it and never criticized Johnson. He would say that he had reservations about our war policy and he would not do things exactly the same way. And that always depressed the kids because they wanted to hear a strong attack. And he knew that, and I think he was really shackled by it enormously because the press would build it up so and he just didn't want to give them any further ammunition."

Kennedy's sensitivity was exacerbated at Berkeley. When asked by a student for his views about the South Vietnamese government, Kennedy responded that while the people of that country didn't want North Vietnamese President Ho Chi Minh, they also certainly didn't want their government's leader, Marshal Nguyen Cao Ky. Kennedy had forgotten (or was unaware) that Ky was at that time meeting with President Johnson at a United States military base in the Pacific. Journalists present that evening seized on the comment, and the *Washington Post*'s David Broder considered leading his coverage of the day's events by claiming that Kennedy was deliberately breaking with Johnson by criticizing Ky in the midst of the presidential conference. Kennedy broke off a reception line to try to convince Broder that he was being misinterpreted, which other journalists then claimed was attempted senatorial intimidation (Broder denied he had been threatened). While Broder's story did not include the side conversation nor characterize Kennedy's remarks as signaling a major split with the administration, the entire incident overshadowed what had been a moving and well-received speech.

Americans for Democratic Action Dinner

PHILADELPHIA, PENNSYLVANIA

FEBRUARY 24, 1967

Kennedy felt his motives were frequently misunderstood by the press. A significant segment of America's young people believed that their actions were misunderstood by their elders. As the gulf between the generations widened, many young people applauded this gap, adopting the slogan "Don't trust anyone over thirty."

Even shared liberalism did not necessarily unite more established groups, like the Americans for Democratic Action, and youthful activists; in particular, students who opposed the administration's course in Vietnam found their motives and even their patriotism questioned. Accordingly, when Kennedy addressed the ADA Convention in 1967, he abandoned the topics he had chosen for previous speeches to the group (shared concerns about civil rights and civil liberties) and attempted to convey to his largely middle-aged audience the point of view of their children. The speech itself reflected the generational division: Kennedy appeared mainly at the request of his friend and former Kennedy administration colleague Arthur Schlesinger Jr., then a fifty-year-old lion of traditional liberalism; the speech was initially drafted primarily by Kennedy's aide Walinsky, then twenty-nine.

When a hundred student body presidents and editors of college newspapers; hundreds of former Peace Corps volunteers; dozens of present Rhodes scholars—when these, the flower of our youth, question the basic premises of the war, they should not and cannot be ignored. Among these protesters, most will serve, if called upon, with courage and responsibility equal to any. But their basic loyalty and devotion does not and cannot obscure the fact of dissent.

These students oppose the war for the same reason that many of you feel anguish: for the brutality and the horror of all wars, and for the particular terror of this one. But for our young people, I suspect, Vietnam is a shock as it cannot be to us. They did not know World War II, or even Korea. And this is a war surrounded by rhetoric they do not understand or accept; these are the children not of the Cold War but of the Thaw. Their memories of Communism are not of Stalin's purges and death camps, not even the terrible revelations of the Twentieth Party Congress, or the streets of Hungary. They see the world as one in which Communist states can be each others' deadliest enemies or even friends of the West, in which Communism is certainly no better, but perhaps no worse, than many other evil and repressive dictatorships all around the world—with which we conclude alliances when that is felt to be in our interest . . .

However the war may seem to us, they see it as one in which the largest and most powerful nation on earth is killing children—they do not care if accidentally—in a remote and insignificant land. We speak of past commitments, of the burden of past mistakes; and they ask why they should now atone for mistakes made before many of them were born, before almost any could vote. They see us spend billions on armaments while poverty and ignorance continue at home; they see us willing to fight a war for freedom in Vietnam but unwilling to fight with one hundredth the money or force or effort to secure freedom in Mississippi or Alabama or the ghettos of the North. And they see, perhaps most disturbing of all, that they are remote from the decisions of policy; that they themselves frequently do not, by the nature of our political system, share in the power of choice on great questions which shape their lives . . .

The nonrecognition of individuality—the sense that no one is listening—is even more pronounced in our politics. Television, newspapers, magazines are a cascade of words, official statements, policies, explanations, and declarations; all flow from the height of government, down to the passive citizen; the young must feel, in their efforts to speak back, like solitary salmon trying to breast Grand Coulee Dam. The words which submerge us, all too often, speak the language of a day irrelevant to our young. And the language of politics is too often insincerity—which we have perhaps too easily accepted but to the young is particularly offensive. George Orwell wrote a generation ago that: "In our time, political speech and writing are largely the defense of the indefensible. Things like the continuation of British rule in India, the Russian purges and deportations, the dropping of the atom bombs on Japan, can indeed be defended, but only by arguments which are too brutal for most people to face, and which do not square with the professed aims of political parties. Thus political language has to consist largely of euphemism, question-begging, and sheer cloudy vagueness."

There is too much truth for comfort in that statement today. And if we add to the insincerity, and the absence of dialogue, the absurdity of a politics in which Byron de la Beckwith[2] can declare as a candidate for lieutenant governor of Mississippi, we can understand why so many of our young people have turned from engagement to disengagement, from politics to passivity, from hope to nihilism, from SDS to LSD.

But it is not enough to understand or to see clearly. Whatever their differences with us, whatever the depth of their dissent, it is vital—for us as much as for them—that our young feel that change is possible; that they will be heard; that the cruelties and follies and injustices of the world will yield, however grudgingly, to the sweat and sacrifice they are so ready to give. If we cannot help open to them this sense of possibility, we will have only ourselves to blame for the disillusionment that will surely come. And more than disillusionment, danger; for we rely on these young people more than we know . . . If we look back with pride at the lives we led, we know above all that we will judge ourselves by the hope and direction we have left behind . . .

And if, when we reach out to them, we are tempted to dismiss their vision as impossible, or their indignation as naive, let us remember, as the poet [Keats] says, that:

None can usurp the height
But those to whom the miseries of the world
Are misery, and will not let them rest.

This speech was lengthened and reworked to become the first chapter of Kennedy's book *To Seek a Newer World,* published later in 1967.

Blacks and the Urban Crisis

By 1967, as he appeared before the Americans for Democratic Action, Keats's epigram clearly fit Kennedy: The miseries of the world would not let him rest. Central to his concerns was the plight of the nation's blacks and the related disintegration of the country's urban centers.

Asked to name Robert Kennedy's most unique characteristic, many friends and commentators during his Senate years (and increasingly in posthumous appraisals) pointed to Kennedy's capacity for personal growth and change. Although Kennedy was more complex than the rich, anti-Communist crusader he was sometimes characterized as being in the 1950s, by 1966 he had clearly changed and was generally regarded as the "tribune of the underclass." And by the end of

Kennedy's life, America's blacks had identified him as their key white champion.

But the roots of Kennedy's black support lay in both the man and his times—in his personal history and the development of the civil rights movement in America. From childhood, he had defended the underdog, and at its most basic, he saw his exposure of the corruption of union officials during the fifties as a vindication of working people betrayed by their leaders and robbed by organized crime. His strong sense of justice didn't change—but both the focus of his energies and the intensity of his identification with the victims did. Kennedy was a man, in Adam Walinsky's view, increasingly moral but less moralistic.

Walinsky pointed to Kennedy's ethnic heritage as formative. Repeatedly in speeches throughout his career, Kennedy remembered that his father grew up in a Boston festooned with signs reading NO IRISH NEED APPLY. Kennedy drew strength from his sense of having descended from exiles and outcasts who had to fight for respect from the power structure, and his forebearers' struggles gave him sympathy for others similarly frozen out.

His brother's assassination also changed him. Schlesinger believed that the events in Dallas "charged sympathy with almost despairing intensity. [Kennedy's] own experience of the waste and cruelty of life gave him access to the sufferings of others. He appeared most surely himself among those whom life had left out."

Critics claimed Kennedy's motivation did not come from a Keatsian discomfort with the "miseries of the world" but from a ruthless desire to "usurp the heights": that he was fashioning a coalition of the dispossessed into a political base to "reclaim" a White House "taken" from the clannish Kennedys by John's assassination. But that political calculation would never have been compelling to a tactician of Kennedy's caliber; clearly, his positions risked alienating the majority of voters, and he was likely to retain the loyalty of minority voters regardless of the intensity of his efforts on their behalf, if only because of the strength of their memory of his martyred brother.

Robert Kennedy also could point with pride to his work for equal justice for blacks, for as attorney general he had coordinated the civil rights policy of his brother's administration and played pivotal roles in the federal intervention to desegregate southern schools and universities, to extend voting rights, and to equalize access to transportation and accommodations. And his involvement went deeper, presaging many of his later policy initiatives.

The President's Committee on Juvenile Delinquency, which Robert Kennedy helped to create and ultimately chaired in his brother's administration, became the first federal agency to work directly with community leadership within poor neighborhoods. His involvement transcended his official role, as when he repeatedly raised funds (and gave substantially himself) and cut bureaucratic impediments to establish playgrounds and recreational programs in black areas of the District of Columbia.

Concerned about the rising and seemingly intractable problem of students dropping out of school, particularly in the inner city, Kennedy spoke repeatedly to school assemblies, cajoling his listeners with visions of a better society that their education would help them create.

Generally, the legislative initiatives of John Kennedy's New Frontier and the legal strategy of the Justice Department under Robert both focused on ending legal discrimination, occasioning a concentration of effort in the South. But unrest was not restricted to the South. In the spring and summer of 1964, what began as civil rights demonstrations and counterdemonstrations had become violent uprisings across a tier of northern cities, including Rochester and New York City. During Kennedy's Senate campaign, he made an early stop at

Rochester (opponent Keating's hometown) and was advised by a local official to avoid all references to blacks and civil rights in a city still shaken by the riot. But before a nearly all-white crowd that filled the Rochester Civic Center to capacity and overflowed into the streets, Kennedy pushed ahead in extemporaneous remarks: "It would have been easier," he said, "not to try to bring this matter up, but there are problems we are all going to have to face. I'm referring to the problems of the races living together."

The audience was stonily silent.

"In the South you can pass legislation to permit a Negro to have an ice cream cone at Howard Johnson's, but you can't pass legislation to automatically give a Negro an education. I believe the community must provide education so there can be jobs."

The applause started weakly in the back, then it grew and spread across the auditorium.

"You have got," said the candidate, "to give Negroes some hope."[3]

In the selection of speeches that follow, Kennedy's evolving views on racial matters and the urban crisis unfold, beginning with a speech in Chicago four months after he was sworn in as senator, where he made clear to his northern audience that equal opportunity was not a regional requirement but a national imperative.

Then a speech from the summer of 1965 is excerpted, as the fires of the decade's first major urban riot still smoldered in Watts.

As 1966 began, Kennedy formulated a comprehensive approach to urban revitalization and racial reconciliation, which he presented in three speeches given on successive days to separate (and disparate) audiences in New York City. Substantial portions of this urban trilogy are included.

The vision outlined in those three January speeches became precise and ready for incorporation into legislation by August, as shown in Kennedy's testimony before a Senate subcommittee, presenting his detailed program for federal intervention in the urban crisis.

By the end of 1966, Kennedy had moved from placing a priority on education and federal government solutions to an emphasis on community empowerment, public/private consortia, and direct job creation. A year of intensive thought and discussion culminated in the launching of the Bedford-Stuyvesant restoration project in Brooklyn, New York, in December 1966; Kennedy's comments at the program's kickoff are provided.

Finally, before the Citizens Union in New York City as 1967 came to a close, Kennedy unified these various threads into a somber and moving Christmas message, which closes this section.

Racial Problems in the North: National Council of Christians and Jews

CHICAGO, ILLINOIS

APRIL 28, 1965

After his surprisingly successful appearances at the universities of Mississippi and Alabama in March 1965, Kennedy turned to northern conditions. He had told an interviewer in 1964 that Dixie residents already "hated [his] guts" for his involvement in southern desegregation but that those legal struggles were "an easy job compared to what we face in the North." He watched with mixed feelings as northern liberals rushed to protests in the South rather than try to improve the conditions in their hometowns. "Why do they go to Selma?" he asked Richard Rovere, a writer. "Why not to 125th Street [in Harlem]?"

Just before resigning as attorney general on September 3, 1964, Kennedy sent President Johnson a memorandum titled "Racial Violence in Urban Centers." In it, Kennedy recognized that inadequate job and housing opportunities could be changed only through years of sustained effort, focused in the cities of the North. He returned to the issue in front of a targeted audience: the National Council of Christians and Jews, at their annual meeting in Chicago on April 28, 1965.

The unfinished business at hand is the most difficult and dangerous that we have ever faced. Today's problems of intolerance are harder than yesterday's; tomorrow's will be harder still.

One reason for this difficulty is that racial intolerance is harder to combat than religious intolerance. Most people, after all, have to be told whether the man they are talking to is Catholic, or Protestant, or Jewish; none need instruction on which are the Negroes or the Puerto Ricans. Most of us can walk on each other's street without arousing comment. If our children go to school together, few of us will know what religion their classmates practice.

But if a Negro walks down a quiet suburban street, or Negro children attend a school, all know it immediately. Simply by being more visible, the Negro is more vulnerable to prejudice . . .

The Negro's heightened visibility makes easier another kind of prejudice and intolerance: prejudice against the poor, intolerance for the unsuccessful. Because Negroes are twice as likely to be unemployed, because their children are three times as likely to be slow in school, even to the point of mental retardation—because of these things, prejudice against Negroes often masquerades as adherence to principles of individual freedom and responsibility. "This is a free country," says the new voice of intolerance. "They have the same chance as anyone else. If they don't take advantage of what we offer, that's their responsibility." And these voices then use the continued extent of Negro poverty, Negro unemployment, and lack of education as an excuse for not doing more . . .

The brutalities of Selma, and its denial of elementary rights of citizenship, were condemned throughout the North; and thousands of white northerners went there to march to Montgomery.

But the many brutalities of the North receive no such attention. I have been in tenements in Harlem in the past several weeks where the smell of rats was so strong that it was difficult to stay there for five minutes, and where children slept with lights turned on their feet to discourage attacks . . .

Thousands do not flock to Harlem to protest these conditions—much less to change them . . .

Action at home requires change in our own way of life. And—in a world already beset by change—to people whose lives are tragically insecure—further unsettling change is unacceptable . . .

It is not enough, in these circumstances, to preach for fair employment, or even to pass a fair employment law. If there are not enough jobs for all, the elimination of Negro unemployment and poverty will be impossible.

Another example: We all know the importance of education for our children, and how severe is the competition for admission to college.

It is not enough to tell a worried parent that prejudice against Negroes is undemocratic; if he fears that desegregation will handicap his child's education, he will fight it almost to the death.

If we wish to achieve peaceful desegregation of the schools—if we wish to improve the quality of education afforded Negro children—we must improve the quality of education throughout our schools, and assure every qualified child the chance for higher education . . .

Education, while vital, is no longer enough . . . We have gone as far as goodwill and even good legislation will take us, and . . . we must now act to bring about changes in the conditions which breed and reinforce intolerance and discrimination.

And if this is true for those who practice discrimination, it is even more true for its victims. If we are to meet our responsibilities—to them, to ourselves, to all our children—we must address ourselves to the difficult and dangerous problems of the urban North . . .

To solve these problems, to ease this frustration, it

is not enough to teach brotherhood in the schools; we must assure that they educate each child to the limit of his capacity. It is not enough, in this technological society, to hire qualified Negroes, nor even try to raise the number that are qualified; we must create new jobs for all that can work, regardless of their level of skill . . .

These are not easy things to do. But the fulfillment of American ideals has never been easy, if only because they are so high.

Three hundred and thirty-four years ago, on a ship sailing to New England, John Winthrop gathered the Puritans on the deck and said, "We must consider that we shall be as a city, set upon a hill, and the eyes of all people will be upon us."

The Puritans were in the middle of the Atlantic when they shared that vision of the city upon the hill. We are still in the middle of our journey. As long as millions of Americans suffer indignity, and punishment, and deprivation because of their color, their poverty, and our inaction, we know that we are only halfway to our goal—only halfway to the city upon the hill, a city in which we can all take pride, a city and a country in which the promises of our Constitution are at last fulfilled for all Americans.

Reflections on the 1965 Watts Riots: State Convention, Independent Order of Odd Fellows

SPRING VALLEY, NEW YORK

AUGUST 18, 1965

Less than four months after he warned his Chicago audience that growing dissatisfaction and frustration in the nation's urban core represented an immediate danger, violence exploded in Watts, then a predominantly black area of Los Angeles. Beatings, looting, arson, and gunfire began in the evening of August 11, 1965, and continued for six days. Thirty-four people died (more Americans than were killed in Vietnam that week) and more than one thousand were injured. Entire blocks were gutted. The nation watched in shock as the carnage unfolded, often in front of television cameras. In suburban neighborhoods far from the violence, fearful whites purchased pistols and shotguns.

Former President Eisenhower linked the events to the permissiveness of the sixties, saying that "the U.S. as a whole has been an atmosphere . . . of lawlessness." His solution was "greater respect for the law." Politicians at all levels echoed Eisenhower, with calls for stronger law enforcement but without much fundamental change.

Kennedy had spent his career enforcing the law, but he was certain that far more was required in addition to law enforcement. The Watts riot also convinced him that his emphasis on education was insufficient. One week after the riot began, he urged a white civic organization in Spring Valley, a suburban community outside of New York City, to examine with greater understanding and compassion conditions faced by blacks and to act with urgency, lest the fires of Watts spread inexorably to other American cities.

I want to speak to you tonight about some of the events of the last week: about the dead and the orphans of the rioting in Los Angeles; about the sick and the distressed of all our urban ghettos; about the hatred and the fear and the brutality we saw in Los Angeles; and about what we can and must do if this cancer is not to spread beyond control.

For it is clear that the riots of the last weekend were no isolated phenomenon, no unlucky chance. They began with a random argument between a drunken driver and a policeman; they could as easily have begun with a fight in a dance hall, as did the riots in Rochester; or with a policeman shooting a boy armed with a knife, as did the riots in New York; or with a fire engine knocking over a lamppost and killing a pedestrian, as did the riots in Chicago.

All these places—Harlem, Watts, South Side—are riots waiting to happen. To look at them is to know the reason why.

First, these are places of poverty. We know that the rate of Negro unemployment is twice the white rate . . . But do we realize also—can we comprehend—that in many census tracts in the core of our cities the unemployment rate may be 25 or 30 or even 40 percent? In the Watts area of Los Angeles, the rate was 34 percent. And in Watts—as in the other areas of this kind—unemployment of young Negroes may go as high as 75 percent.

Our society—all our values, our views of each other, and our own self-esteem; the contribution we can make to ourselves, our families, and the community around us; all these things are built on the work we do. But too many of the inhabitants of these areas are without the purpose, the satisfaction, or the dignity that we find in our work. And lack of work means lack of money—and living in overcrowded, rat-infested housing, or even renting cars for a night to have a place to sleep, as many people do in the Watts area.

More important, these are places of blighted hopes and disappointment. Their inhabitants came north, in the words of one of them, feeling "as the Pilgrims must have felt when they were coming to America."

But someone had neglected to tell the folks down home about one of the most important aspects of the promised land: It was a slum ghetto. There was a tremendous difference in the way life was lived up north. There were too many people full of hate and bitterness crowded into a dirty, stinky, uncared-for, closet-size section of a great city.

The children of these disillusioned colored pioneers inherited the total lot of their parents: the disappointments, the anger. To add to their misery, they had little hope of deliverance. For where does one run to when he's already in the promised land?

Their disappointment is all the keener because of the prosperity and the affluence all around them . . .

Disappointment and disillusionment have come also from our actions and our promises. We say to the young, for example, "Stay in school, learn and study and sacrifice, and you will be rewarded the rest of your life." But a Negro youth who finishes high school is more likely to be unemployed than a white youth who drops out of school—and is more likely to find only menial work at lower pay . . .

After all, we are very proud of the fact that we had a revolution and overthrew a government because we were taxed without representation. I think there is no doubt that if Washington or Jefferson or Adams were

Negroes in a northern city today, they would be in the forefront of the effort to change the conditions under which Negroes live in our society.

But we have been strangely insensitive to the problems of the northern Negro. During the Birmingham crisis in 1963, I met with many northern leaders: businessmen, newspaper publishers, civic officials. I told them that they would soon face problems even more difficult in their own communities. To a man, they denied that any such problem could arise in the North.

And now—even after the crisis has come—we continue to be surprised by how difficult the problems are to solve. In the last four years, the Negro has made great progress; and the Civil Rights Act of 1964 and the Voting Rights Act of 1965 are rightfully regarded as achievements of which we can all be proud. But as we are learning now, it is one thing to assure a man the legal right to eat in a restaurant; it is another thing to assure that he can earn the money to eat there . . .

The southern civil rights movement is different from northern problems in more serious ways. In the South, the movement has been strongly led and relatively disciplined. Many of the leaders—men like Dr. King, Dr. Abernathy, the Reverend Shuttleworth—have been ministers. All have preached and practiced nonviolence.

But unfortunately—and dangerously—northern problems are the problems of everyday living, in jobs and housing and education. They affect too many people, too directly, for involvement to be restricted to those with the patience, the discipline, and the inclination to practice nonviolence. The army of the resentful and desperate is larger in the North than in the South, but it is an army without generals, without captains, almost without sergeants. Civil rights leaders cannot, with sit-ins, change the fact that adults are illiterate. Marches do not create jobs for their children. So demagogues have often usurped the positions of leadership, each striving to outdo the other in promise or threat, offering unreal hopes and dangerous hate. The result is that during and after every riot, city and police officials have pleaded for leaders, for representatives of the Negroes who could negotiate an end to the violence and establish a channel of communication to the rioters. Each time, Negro civil rights leaders have frankly acknowledged their inability to lead the mobs or the neighborhoods from which the mobs have come.

Partly, of course, the absence of leaders capable of stopping violence is due to the fact that many of the rioters are simply hoodlums, with nothing more in mind than booty and the thrill of defying the law . . .

There is no question that hoodlums should be treated as such.

The time to consider the causes of hoodlumism is before or after—not during—a riot. Poverty and education programs do not stop bullets from killing or fire from burning. Any riot should and must be put down with the use of sufficient force to stop it quickly, with a minimum of damage to lives and property.

At the same time, we must realize that force by itself is no solution . . .

And one harsh fact with which we must learn to live is that just saying "obey the law" is not going to work. The law to us is a friend, which preserves our property and our personal safety. But for the Negro, law means something different. Law for the Negro in the South has meant beatings and degradation and official discrimination; law has been his oppressor and his enemy. The Negro who has moved north with this heritage has not found in law the same oppression it meant in the South. But neither has he found a friend and protector. We have a long way to go before law means the same thing to Negroes as it does to us. The laws do not protect them from paying too much money for inferior goods, from having their furniture illegally repossessed. The law does not protect them from having to keep lights turned on the feet of children at night, to keep them from being gnawed by rats. The law does not fully protect their lives—their dignity—or encourage their hope and trust for the future . . .

The first step is to move beyond thinking about this as a "Negro problem." The difficulties these people face are far greater because of the color of their skin; but the problems themselves are as various as ours are. The problems of a Negro mother without a husband are not the same problems faced by an unemployed Negro teenager. The difficulties of a youth addicted to narcotics are related to, but are not the same as, the problems this youth creates for the Negro family trying to build a decent life in the slum . . .

All of these people are handicapped; all need our help if we are to right what President Johnson called "the American failure, the one huge wrong of [our] nation." But if our help is to be meaningful, it must be directed at them as people—not as a single class labeled *Negro.*

The second broad step we must take is to bring these problems into the political process—to make them the subject of public action . . . Only when these problems are being dealt with in the political process can we expect to channel people's frustration and resentment and insecurity into constructive action programs. The drive for Negro voting in the South has been peaceful in large part because peaceful protest found in the federal government an audience and an authority capable of taking action and willing to do so. High unemployment among Negroes in the North has resulted in more riots than peaceful protests in large part because government has not had the tools to directly affect the wide margins between Negro and white unemployment rates . . .

A way must be found to stop this waste of human resources and the resulting financial drain on the rest of the community. We cannot afford to continue, year after year, the increases in welfare costs which result when a substantial segment of the population becomes permanently unemployable. We cannot afford the loss of the tax revenue we would receive if these people were jobholders. We cannot afford the extra police costs that slums bring. Our slums are too expensive; we cannot afford them.

There will be much to be done at the state and local level as well, especially in giving to the poor a voice in the decisions which affect their daily lives. It is only by affording them such a voice that we offer them an alternative to the streets. It is only by inviting their active participation that we can help them develop leaders who make the difference between a political force—with which we can deal—and a headless mob—with which no one wants to deal.

And more leadership will have to come from the Negroes themselves. Admitting all the difficulties of dealing with northern problems, it is still disappointing and dangerous that Negro leaders of stature have devoted their major efforts to the South—which is important but is only one part of a larger problem. And too many Negroes who have succeeded in climbing the ladder of education and well-being have failed to extend their hand and help to their fellows on the rungs below. We will demand much of ourselves in the years ahead. We should demand as much from the many Negroes who already share our advantages.

There should be no illusions, however, that any or all of these steps will usher in the millennium. Poor people and Negro leaders will make mistakes; there will be graft and waste. We will be tempted to run these programs ourselves for their benefit, until we recall the wisdom of Winston Churchill, who said that democracy is the worst form of government ever invented—except for all the others that have ever been tried.

In emphasizing action on the root causes of discontent rather than merely condemning violence, Kennedy was deliberately out of step with public opinion. The scant press attention paid to the speech exaggerated the differences between popular hardline attitudes and Kennedy's position, with some suggesting that Kennedy believed that the law was the enemy of black Americans.

Clearly, the August speech didn't represent a reversal of Kennedy's respect for the rule of law. He also continued to champion greater compensation, training, and respect for police officers, including proposing draft deferment status (such as then extended to college students) for young men who committed to a three-year tour as cops.[4]

In November 1965, Kennedy and aide Peter Edelman toured Watts, walking in neighborhoods still devastated after the riots of the previous August. Edelman remembered: "It was like seeing a war-torn area in one's own country for the first time." At one point, the two were

> *standing on a street corner in Watts . . . and asking a man—he was a perfectly healthy-looking man—what the problem was. The man said to [Kennedy], "Frustration." Kennedy said, "Excuse me? . . . What do you mean?"*
>
> *. . . He said, "Frustration . . . Well, man, when*

> *you is over fifty and you is black, you can't get a job nohow."*

That view was echoed by urban blacks Kennedy spoke to elsewhere, including in his own state. Edelman remembered Kennedy hearing "over and over again, 'Well, why the hell should I stay in school? There's no job for me when I get out.'"

While in Watts, Kennedy and Edelman visited various job-training efforts springing up in the community in the wake of the riots. Although encouraged by the spirit that inspired them, Kennedy knew how fragile such programs were and how large was the need. He returned from his trip determined to focus on job creation as a centerpiece of a coordinated attack on the problems of the cities, and he asked his staff to devise a detailed plan for the 1966 congressional session.

A Program for the Urban Crisis: A Series of Three Speeches

Federation of Jewish Philanthropies of New York
Borough President's Conference of Community Leaders
United Auto Workers Regional Conference

NEW YORK, NEW YORK

JANUARY 20, 21, AND 22, 1966

One summer day after the Spring Valley speech in 1965, Kennedy invited Walinsky, Edelman, and his administrative assistant, Joe Dolan, to Hickory Hill, his Virginia home. The poolside discussion centered around fashioning an effective program to move the nation beyond both the majority's reaction to Watts and the conditions bred there. Kennedy knew that education, even if significantly reformed, would not alone suffice. Walinsky, thinking of a draft article he had prepared the year previous (while still at the Justice Department), argued for a massive urban reconstruction program, placing economic development at the fulcrum of solving the urban crisis (and moving toward racial equality).

Kennedy was interested but unconvinced. He asked Walinsky to develop his concept; the aide remembered that Kennedy "was damn sure going to know what he was proposing [in detail] before he did it, and he wasn't afraid of saying, well, it's going to cost 10, 15, 20, whatever number of billions of dollars; but he wanted to be able to say where they were coming from, and where they were going to go, and what they were going to do, and what the programs were going to be."

Meanwhile, Edelman was working with Burke Marshall (formerly Kennedy's principal Justice Department deputy on civil rights, who had become a corporate lawyer) on proposals to assure equal justice and opportunity. The result was a lengthy Edelman memo to Kennedy in November 1965, which the latter read on his return from Latin America. Edelman chiefly concentrated on extending legal rights, particularly in the South, but the memo concluded with a brief acknowledgment that an alternative focus could be on the structural issues (economic opportunity, housing, education, and the like) facing the North, topics that had been the subject of the Spring Valley speech in August. Kennedy told Edelman to emphasize these northern, urban issues.

The two legislative aides spent hours discussing the content and structure of a major address, to accomplish Kennedy's aim of launching a thorough urban effort. Their contact with their boss was regular, but often under chaotic conditions: Edelman remembered one crucial policy discussion taking place in a room at journalist Rowland Evans's home, where Kennedy was joining Washington's elite for a dinner party. Kennedy directed Walinsky and Edelman to reflect his views in a draft of a multipart speech that he could deliver on successive days in January, utilizing previously scheduled events.

The initial structure of considering education, housing, and employment as separate legs of a triad of speeches proved unwieldy. The revised approach was, Walinsky recalled, "the best stuff Kennedy ever did on urban issues. The first speech [initially drafted by Edelman] concentrated on how best to pursue the goal of integration; the second [from Walinsky], recognizing that integration would take a long time, addressed what you could do in the ghetto now; and the third [also initially Walinsky's work] described what the bargain would be for working whites, to enlist their support."

Reviewing the draft of the trilogy at Kennedy's request, Burke Marshall commented: "Well, it's very radical. But then, the challenges are radical." Kennedy's office stimulated media interest with advance press advisories explaining the tripartite structure of the speeches and calling the effort a major policy address. Kennedy began at a luncheon of the Federation of Jewish Philanthropies of New York, held at the Americana Hotel on January 20, 1966.

My remarks today are the first in a series of three speeches about one major aspect of the unfinished business that is ahead: the quality of life for the Negro in the urban areas of the North.

I do not mean to downgrade the problems that remain in the South. But my purpose today—and in the speeches I shall deliver tomorrow and Saturday—is to emphasize the magnitude of the problem in the North, and to suggest that our purpose to help the northern Negro must pervade every plan we make for the future of our cities. In the course of the last generation, the face of the American city has changed almost completely. Millions of Negroes, impelled by hopes of job opportunities and a better life, have poured into the great cities of the North and West. Their arrival, and the simultaneous flight of millions of whites to the suburbs, has created a situation of segregation unparalleled in our country's history . . .

And the thoroughgoing segregation of the urban Negro is actually the less serious part of the story. The worst part is that the ghetto is also a slum.

From his birth to his death the urban Negro is a second-class citizen.

Kennedy then cited a wealth of statistical proof of the comparative disadvantage of blacks in jobs, education, housing, health, and other aspects of the quality of life.

. . . For three hundred years the Negro has been a nation apart, a people governed by a repression that has been softened to the point where it is now only a massive indifference. The Watts riots were as much a revolt against official indifference, an explosion of frustration at the inability to communicate and participate, as they were an uprising about inferior jobs and education and housing. What exploded in Watts is what lies beneath the surface . . .

We can expect continuing explosions like Watts, continuing crises in the management of our cities, and, worst of all, a continuing second-class status for a large group of American citizens. Clearly the present pace is unsatisfactory . . .

What, then, are the specific elements of our course of action? Ultimately, we must succeed in wiping out the huge central city ghettos. By this I do not mean that the outcome will be racial balance in every urban and suburban neighborhood. Many Negroes, given a completely free choice, will choose to live in predominantly Negro neighborhoods, just as members of other racial groups have chosen in the past to live predominantly among their own kinsmen.

The important thing is that the Negro must have freedom of choice. What we must achieve is freedom for him to move if he wants to and where he wants to . . .

If we can break down the massive housing segregation of the ghetto, we can break down the other forms of segregation which it has caused. The ghetto, for example, makes it practically impossible to achieve meaningful racial balance in the schools. Even if there were enough whites left in the city to make a balance, the physical distances involved in transporting children, particularly in a large city, might be insuperable . . .

Obviously we cannot wipe out the ghetto overnight. Were this our sole goal, we could succeed only in writing off a whole generation of present ghetto residents who would never live to see its fulfillment.

At the same time, therefore, we must devote our attention to improving living conditions and rebuilding the present Negro areas, to giving their residents new job skills and jobs to go with them, to improving the education of their children, to providing new cultural interests for those who live there.

Many have said that the goals of improving living conditions in Negro areas and of dispersing the ghetto are mutually exclusive. I disagree most emphatically. In fact, the two goals are interdependent . . .

What can we do now to begin reversing the concentration of the Negro in the central city? . . .

First, we must evaluate existing federal housing programs in light of the goal of desegregation. It will turn out, of course, that we have been doing things just about backwards for a long time.

Public housing has been a significant force in perpetuating segregation. It has been built, on the whole, as large projects in ghetto areas, and it is all too clear that this has not been wholly accidental . . .

Second, we can act at the federal level to break down the barriers which the suburbs present to the relatively few Negroes who can afford now to buy a home . . . [and] to outlaw discrimination in the sale or rental of housing . . .

Third, we can begin making efforts in other areas to complement our efforts in the housing area.

We might, for example, consider special federal aid to suburban schools which take in slum children . . .

Finally, it would be invaluable to desegregation if we established, with federal assistance, well-publicized advisory agencies to tell new arrivals to the city of available places to live, available jobs, and so on . . .

It is the inevitable erosion of the spirit which isolation [of the ghetto] has brought that we seek to counteract . . . Our ultimate purpose is to assure that every American comes to know the full meaning of the truths that we held to be self-evident for the rest of America almost 190 years ago.

Regardless of the preparation and substantive content, the speech was, Edelman remembered, "an utter, total bomb. Walking into a chicken à la king lunch . . . and giving this very long, very serious, very heavy speech—really threatening all of them because they were all the people who he was talking about—it was just a bomb, and he was really very irritated with me afterwards."

The next day, when Kennedy addressed mostly black community leaders at a conference called by the presidents of New York City's five boroughs, the audience was certainly easier and the reaction far more favorable.

Since the dawn of their freedom a century ago, Negro Americans have been advised to "cast down your bucket where you are."

But those who offered this advice too often did not bother to look at whether its recipient was standing by a river of opportunity—or in the midst of a desert from which his bucket could bring back only the sand of poverty and ignorance and want . . .

Clearly, the most important problem in Harlem is education of every kind. Fathers must learn job skills, and mothers how to buy food economically; students must learn to read, and little children how to speak; and teachers must learn how to teach and employers how to hire.

But our educational efforts have thus far not been sufficient . . .

Fundamentally, better schools do not automatically produce better or more dedicated students. For perhaps the greatest barrier to education in Harlem is simply a lack of hope, a lack of belief that education is meaningful to a Negro in the city of New York.

The proof can be found in one fact: that the most important factor in the dropout rate is not principally the state of the student's family, nor his earlier schooling, nor whether he attended preschool classes or had a guidance counselor. The main determinant of the dropout rate is the opportunity for employment in the area.

This, after all, should be no surprise. Children whose fathers are without work for months or years on end are not likely to learn the value of work in the school or elsewhere. High school seniors who see last year's graduates standing on street corners, or working part-time in menial jobs, are not likely to be impressed with the value of the last year's schooling.

And the effects of the shortage of meaningful employment are reinforced by a welfare structure which is frequently destructive both of individuals and of the community in which they live . . .

So if we are to break out of this cycle—if our educational programs are to work—we must move immediately to provide jobs for all those willing and able to work . . . Arguments over whether unemployment can best be improved by job training or general prosperity become meaningless. Neither has been adequate. Something more is needed.

At this point let us make clear what that something more is not.

It is not a massive extension of welfare services

or a new profusion of guidance counselors and psychiatrists, whether on a block, neighborhood, or other basis . . .

I turn now, therefore, to a more realistic program for jobs—a program for all our Harlems—a program for the United States.

Action.

Let us, as a beginning, stop thinking of the people of Harlem—the unemployed, the dropouts, those on welfare, and those who work for less than the minimum wage—as liabilities, idle hands for whom some sort of occupation must be found. Let us think of them instead as a valuable resource, as people whose work can make a significant contribution to themselves, their families, and the nation.

Now ask if there are jobs to be done . . . In fact, the inventory is almost infinite: parks and playgrounds to be built, the beaches to be renovated, the subways to be refurbished. If we begin—as the president said we will—to meet the pressing needs, there will be jobs enough for all our people.

But let us not make the mistake of regarding these just as jobs; and let us not erect buildings for their own sake. In any program of rebuilding now begun, therefore, I urge the following:

First: Priority in employment on these projects

should go to residents of the areas in which they are undertaken.

For this is man's work—work which is dignified, which is hard and exacting, which is at the same time rewarding to the man who does it and rewarding to the community around him. Much of it is work which can be done by unskilled workers, who now have the most difficult time finding jobs; but in such a program there would be jobs of all kinds, including those requiring administrative and managerial skills . . .

Second: Public and private training programs should concentrate their funds and their efforts in on-the-job training on these projects . . .

Third: Our conventional educational system should be directly integrated with the rebuilding effort . . .

Any high school student who so desired—whether for financial or other reasons—could be allowed to leave school to work on such a project. The schools would maintain jurisdiction over these students; and they would, as a condition of employment, be required to continue schooling at least part-time until the requirements for graduation were met . . .

Fourth: The rebuilding should be consciously directed at the creation of communities: the building of neighborhoods in which residents can take pride, neighborhoods in which they have a stake, neighborhoods in which physical surroundings help the residents to create the functioning community which must be our goal. We should, for example, make provision for condominium ownership of low-income apartments. At another level, we should engage in as much rehabilitation as possible, saving all of the old that is economical and sound. We should build in stores and workshops and play space. And the planning of the neighborhoods should, from the outset, involve the people of the areas affected.

Fifth: Present social service programs, particularly welfare, should be integrated with the rebuilding effort . . .

Sixth: Using the building program as a base, occupational opportunities and training should be opened up in all related ways. As building takes place, for example, some should learn and then operate building-supplies businesses; small furniture-manufacturing establishments; or restaurants in which the workers can eat . . .

Seventh: An essential component of any program for regeneration of the ghetto will be the active participation of the business community in every aspect of the program, in a partnership of shared costs and effort with government . . .

Eighth: Equally essential is the full participation of private groups, especially of labor unions and of universities . . .

What is called for, in short, is a total effort at regeneration: an effort to mobilize the skills and resources of the entire society, including above all the latent skills and resources of the people of the ghetto themselves, in the solution of our urban dilemma.

This is a lengthy list, yet even so it is not complete . . .

Finally, if our plans are to be serious, not merely an exercise in planning, there is the question of cost. I do not believe that cost to be prohibitive; much is already available under existing programs . . .

This is not to say that all the cost can be met out of the present budget; it cannot. But a start can be made, in some of our great cities—a start which will teach us the techniques of such a massive effort . . .

And any costs of this program will be to a substantial degree offset by lower welfare costs . . .

But the greatest returns of the program I have outlined would be returns in human spirit: in lessened dependency, in lower delinquency and crime, in more beautified cities, and children stronger and healthier in every way; and these returns are beyond our capacity to measure . . .

None of this can happen unless you will it to happen, and unless you can and do make the hard and sometimes unpopular decisions which come with responsibility.

It is easy and popular, for example, to attribute our housing problems to the greed of landlords and the indifference of the city, and to seek ever more drastic penalties for failure to make needed repairs, or even more drastic protests against incomplete enforcement.

Certainly there are many landlords who have profited shamelessly from the people of Harlem.

But to seek the solution for our housing problems in ever more rigorous code enforcement is irrelevant. For after a generation of trying, it should be clear to all that we simply do not have the legal and administrative resources to chase every landlord in Harlem through the courts and boards whenever the heat fails or a window is broken.

What is needed is to put these buildings in the hands of the people who do want to keep them up. But we have not done what we could toward this end . . .

So it is you—leaders of the Negro American community—who know what must be done better than white Americans can ever know—you must take the lead; you must take the first steps, using what is available, and showing what is needed but not available . . .

> The dream of liberty and equality set out in our Constitution 190 years ago [was] never completely fulfilled, yet [is] always alive, still asserting itself with ever-growing insistency. To work to bring it closer to fruition is the highest task in which we can engage.

Not surprisingly, Kennedy's program received no complaints from an audience composed of the very community leaders he proposed to empower. Far more difficult was the task of enlisting the acceptance and, hopefully, assistance of working whites in massive federal efforts to help others and to rebuild the cities these blue-collar whites had left or were leaving.

Organized labor offered a potential but ticklish challenge. Union support was an important part of Kennedy's political base in New York and labor was a vital part of the national Democratic coalition, with nearly a third of the delegates to the Democratic National Convention being either union members or those who owed their allegiance to organized labor. Union involvement in liberal issues was long-standing.

But many unions resisted racial changes in their shops. Construction trades, for example, frequently denied blacks apprenticeships and, thereby, access to union work and union wages. And blue-collar whites felt most threatened by the racial and economic advancement of blacks: terrified of those they viewed as sources of riots and crime, concerned that interracial neighborhoods meant lower property values, and afraid that more blacks in the work force meant more white layoffs, not to mention resistant to paying higher taxes to fund it all.

The United Auto Workers was an integrated union, although the top leadership was white. Kennedy chose a New York–area UAW conference to test an appeal he would echo for the remainder of his life: The majority could advance its quality of life by ensuring that the poorer minority had both life's necessities and the full opportunities for success. As Walinsky recalled later, "A true poverty program has to offer the middle class a new lifestyle in exchange for the things that it would give to the poor. And [such a program] must deal not only—or even primarily—with pockets of economic poverty, but rather with a poverty of satisfaction [and] purpose . . . The larger social fabric . . . [and inspiring all to be a part of it, is] what leadership in a democracy consists of."

Tonight I want to speak with you not about labor unions, but about the common future of those who labor; about what has too often happened to the great American dream, and about how we can bring new opportunities to millions of Americans.

The Americans of whom I speak, however, are not the poor, or the Negro, or the many victims of our slums.

The Americans of whom I speak share but one characteristic with the poor: that the opportunities available to them are not as full and free as they could and should be. I speak of all Americans.

We are now engaged in a great experiment: an attempt to bring new education, new training, new jobs to the poor and the distressed . . . an attempt to bring new comfort to the elderly and new fulfillment to the young.

In these attempts, we are exercising our imaginations and our will, and the power of government to initiate change, to enrich the lives of one or another group of our people.

But we have not exercised that same imagination and will, or the same ability of government to initiate and stimulate action, on behalf of Americans who are not poor, not cut off from the twentieth century, not handicapped by racial discrimination or age or youth.

The question will be: Why should we? What help do these more fortunate Americans need—in a time of record gross national product, of increasing prosperity throughout the society?

The answer can be found by examining a series of current problems . . . A . . . major problem we now face is a mismatch between available jobs and people without work. The greatest demand for new workers, now and for the next ten years, will be in the professional, technical, and highly skilled fields . . . The greater bulk of our unemployed, by contrast, are those with the least education, the least training, the lowest skills . . .

It would seem to make better sense to upgrade all of our workers . . .

The time has come when we must open the channels of opportunity—the chance for new labor and new lives—for all our citizens.

To this end, I propose:

First: We should now begin to study ways in which workers presently in the labor force could be assisted and encouraged to resume their education, continuing through college . . .

A program of . . . extending opportunity for higher education might well be directed, at least in part, to meeting our urgent needs for public-service workers. For example, the Naval Reserve Officers' Training Course now pays students full tuition, room, board, and incidental expenses in return for four years' guaranteed service after graduation. We might extend scholarships to older students on a similar basis, perhaps requiring three or four years' work in Appalachia or Harlem, or in technical assistance abroad, in return . . .

Second: State and local educational systems should engage in similar studies of ways to further the education of adults who wish to return to school . . .

Third: Federal and state job-training programs should be made available, as resources permit, to persons now employed . . .

Fourth: An expanded and more vital U.S. Employment Service could act as a nationwide clearinghouse of jobs and job-training opportunities . . .

Fifth: Any effort to upgrade skills throughout the economy will require close partnership of business, labor, and government . . .

Sixth: If we are to allow new mobility to our people, we will have to assure them that rights they have earned—equities built up in pension funds—are not lost . . . We should now, therefore, actively explore the possibilities of progressively reducing the vesting period and insuring transferability of pension rights . . .

Seventh: The benefits of further education and training should not be thought of as only for younger workers . . . We should make every attempt to use [the] full potential [of retired workers], particularly in community-service work; perhaps part-time training for it could take place during the last several years before retirement.

The foregoing is no exhaustive list. By its very incompleteness it demonstrates that we are only at the beginning of a beginning in thinking about opening opportunities for all. More thinking must begin now. For the pace of change—in our economy, in our politics, in our whole society—can only accelerate.

Change is chance—which, as Pasteur said, "favors the mind that is prepared." Let us now then prepare the minds of all our people—and the tides of chance and change will favor all of us.

The UAW gave Kennedy a warm reception. The next day, he met with Walinsky and Edelman. "Now listen," the senator said, "I don't just want to talk about this; I want to do something about it . . . Now we've got to do it." He gave his aides two principal tasks: to flesh out the programmatic details of the federal role, to become legislation; and to search for a venue in New York to set up an innovative demonstration of the unified, community-based effort he was suggesting. This work coalesced in Kennedy's Senate testimony in August 1966 and, after nearly twelve months of intensive discussions and tinkering, in the launching of the Bedford-Stuyvesant project in December of that year.

Solving the Urban Crisis: Testimony Before the Subcommittee on Executive Reorganization

WASHINGTON, D.C.

AUGUST 15, 1966

As Kennedy became convinced that far greater federal leadership would be required to coordinate revitalization of American cities, the costs of the United States' involvement in Vietnam began to impact the domestic budget significantly. When President Johnson proposed cutbacks in education and a variety of poverty programs, Kennedy was outraged. In an April 19, 1966, speech in Ellenville, New York (at the edge of the Catskills), he attacked the cuts, but not directly their presidential sponsor, as irresponsibly shortsighted. The press played up the speech as demonstrating further rifts between two of the country's most powerful politicians.

With budget constraints looming, the future direction of the War on Poverty[5] became the focus of an important series of hearings in August 1966 held by the Subcommittee on Executive Reorganization of the Senate Committee on Government Operations. Kennedy took the somewhat unusual step of testifying before a subcommittee of which he was a member, in hopes that the sessions that followed would be stimulated by (and supportive of) both his lengthy context-setting exposition of urban problems and the detailed programmatic solutions that he and his staff had been honing since January's trilogy of speeches on the urban crisis. He appeared on August 15 in a Capitol hearing room.

Each of our cities is now the seat of nearly all the problems of American life: poverty and race hatred, interrupted education and stunted lives, and the other ills of the new urban nation—congestion and the filth, danger, and purposelessness which afflict all but the very rich and the very lucky. To speak of the urban condition, therefore, is to speak of the condition of American life. To improve the cities means to improve the life of the American people . . .

The city is . . . a place where men should be able to live in dignity and security and harmony, where the great achievements of modern civilization and the ageless pleasures afforded by natural beauty should be available to all.

If this is what we want . . . we will need more than poverty programs, housing programs, and employment programs, although we will need all of these. We will need an outpouring of imagination, ingenuity, discipline, and hard work unmatched since the first adventurers set out to conquer the wilderness . . .

One great problem is sheer growth—growth which crowds people into slums, thrusts suburbs out over the countryside, burdens to the breaking point all our old ways of thought and action, our systems of transport and water supply and education, and our means of raising money to finance these vital services.

A second is destruction of the physical environment, stripping people of contact with sun and fresh air, clean rivers, grass and trees—condemning them to a life among stone and concrete, neon lights and an endless flow of automobiles. This happens not only in the central city but in the very suburbs where people once fled to find nature . . .

A third is the increasing difficulty of transportation, adding concealed, unpaid hours to the workweek; removing men from the social and cultural amenities that are the heart of the city; sending destructive swarms of automobiles across the city, leaving behind them a band of concrete and a poisoned atmosphere . . .

A fourth destructive force is the concentrated poverty and racial tension of the urban ghetto, a problem so vast that the barest recital of its symptoms is profoundly shocking . . .

Fifth is both cause and consequence of all the rest. It is the destruction of the sense, and often the fact, of community, of human dialogue, the thousand invisible strands of common experience and purpose, affection and respect, which tie men to their fellows. It is expressed in such words as community, neighborhood, civic pride, security among others, and a sense of one's own human significance in the accepted association and companionship of others . . .

Nations or great cities are too huge to provide the values of community. Community demands a place where people can see and know each other, where children can play and adults work together and join in the pleasures and responsibilities of the place where they live. The whole history of the human race, until today, has been the history of community. Yet this is disappearing, and disappearing at a time when its sustaining strength is badly needed. For other values which once gave strength for the daily battle of life are also being eroded.

The widening gap between the experience of the generations in a rapidly changing world has weakened the ties of family; children grow up in a world of experience and culture their parents never knew. The world beyond the neighborhood has become more impersonal and abstract. Industry and great cities, conflicts between nations and the conquests of science move relentlessly forward, seemingly beyond the reach of individual control or even understanding. It is in this very period that the cities, in their tumbling spread, are obliterating neighborhoods and precincts. Housing units go up, but there is no place for people to walk, for women and their children to meet, for common activities. The place of work is far away through blackened tunnels or over impersonal highways. The doctor and lawyer and government official is often somewhere else and hardly known. In far too many places—in pleasant suburbs as well as city streets—the home is a place to sleep and eat and watch television; but the community is not where we live. We live in many places and so we live nowhere.

Long ago, de Tocqueville foresaw the fate of people without community: "Each of them living apart is a stranger to the fate of all the rest—his children and his private friends constitute to him the whole of mankind; as for the rest of his fellow citizens, he is close to them, but he sees them not; he touches them but he feels them not . . . he may be said at any rate to have lost his country." To the extent this is happening it is the gravest ill of all. For loneliness breeds futility and desperation, and thus it cripples the life of each man and menaces the life of all his fellows.

But of all our problems, the most immediate and pressing, the one which threatens to paralyze our very capacity to act, to obliterate our vision of the future, is the plight of the Negro of the center city. For this plight, and the riots which are its product and symptom, threaten to divide Americans for generations to come, to add to the ever-present difficulties of race and class the bitter legacy of violence and destruction and fear.

The riots which have taken place, and the riots which we know may all too easily take place in the future, are therefore an intolerable threat to the most essential interests of every American, black or white, to the mind's peace and the body's safety and the community's order, to all that makes life worthwhile. None of us should look at this violence as anything but destructive of self, community, and nation. But we should not delude ourselves. The riots are not crises which can be resolved as suddenly as they arose. They are a condition which has been with us for one hundred years and will be with us for many years more. We can deal with the crises without dealing with the underlying condition—just as we can give novocaine to a man with a broken arm, without setting that arm in a sling; but the end result will only be more pain, pain beyond temporary relief, and permanent crippling of our urban society . . .

It is clear that our present policies have been directed to particular aspects of our problems, and have often ignored or even harmed our larger purposes. For example, federal housing and highway programs have accelerated the move of the middle-income families and businesses to the suburbs, while virtually ignoring the cities' needs for new revenue and declining tax base . . .

Our housing projects were built largely without either reference or relevance to the underlying problems of poverty, unemployment, social disorganization, and alienation which caused people to need assistance in the first place. Too many of the projects, as a result, become jungles—places of despair and danger for their residents and for the cities they were designed to save . . .

No single program, no attempted solution or any single element of the problem, can be *the* answer.

In recent years, education has come to be regarded as the answer . . . But past efforts to improve life conditions simply by the expenditure of more money on education have not been notably successful . . .

Education has failed to motivate many of our young people because of what they could see around them: the sharply restricted opportunities open to the people of the ghetto, whatever their education . . .

We know the importance of strong families to development; we know that financial security is important for family stability and that there is strength in the father's earning power. But in dealing with Negro families, we have too often penalized them for staying together. As Richard Cloward has said: "Men for whom there are no jobs will nevertheless mate like other men, but they are not so likely to marry. Our society has preferred to deal with the resulting female-headed families not by putting the men to work but by placing the unwed mothers and children on public welfare—substituting check-writing machines for male wage-earners. By this means we have robbed men of manhood, women of husbands, and children of fathers. To create a stable monogamous family, we need to provide men (especially Negro men) with the opportunity to be men, and that involves enabling them to perform occupationally."

And here we come to an aspect of our cities' problems almost untouched by federal actions: the unemployment crisis of the Negro ghetto . . .

The crisis in Negro unemployment, therefore, is significant far beyond its economic effects, devastating as those are. For it is both measure and cause of the extent to which the Negro lives apart . . .

Unemployment is having nothing to do, which means having nothing to do with the rest of us.

It is a shocking fact—but it is a fact nonetheless—that . . . our . . . whole array of government computers, which threaten to compile on some reel of tape every bit of information ever recorded on the people in this room—this system nowhere records the names or faces or identities of a million Negro men.

Seventeen percent of Negro teenagers, 13 percent of men in the prime working age of the thirties, are uncounted in our unemployment statistics, our housing statistics, simply drifting about our cities, living without families, as if they were of no greater concern to our daily lives than so many sparrows or spent matches.

Some are "found" in later life, when they may settle down. Some reappear in our statistics only at death. Others remind us of their presence when we read of rising crime rates. And some, undoubtedly, became visible in the riots . . . I stress employment here for the following reasons:

First, it is the most direct and embarrassing—and therefore the most important—of our failures . . .

Second, employment is the only true long-run solution. Only if Negroes achieve full and equal employment will they be able to support themselves and their families, become active citizens and not passive objects of our action, become contributing members and not recipients of our charity . . .

Third, there are government programs which seem at least to have some promise of ameliorating, if not solving, some of the other problems of the Negro and the city. But no government program now operating gives any substantial promise of meeting the problem of Negro unemployment in the ghetto . . .

[We] must attack these problems within a framework that coordinates action on the four central elements: employment, education, housing, and a sense of community.

This is not to say that other problems and programs are not important: questions of police relations, recreation, health and other services, and the thousands of other factors that make life bearable or a thing of joy. It is to say that these other questions can only be properly dealt with in concert with action on the major problems. A police force, for example, can exert every possible effort, and imagination, and will, to better relations with the community. But it still must enforce the law. And if the conditions of the ghetto produce stealing, for which people must be arrested, or nonpayment of rent, for which people must be evicted, even if they have no place to go, then the police will inevitably bear the brunt of the ghetto's resentment at the conditions which the police, through no fault of their own, enforce . . .

Kennedy then outlined his jobs program, described in his January 21, 1966, speech in New York.

To bring the people of the ghetto into full participation in the economy, which is the lifeblood of America, it will be necessary to create new institutions of initiative and action, responding directly to the needs and wishes of these people themselves. This program will require government assistance, just as nearly all American growth has depended on some government assistance and support. But it cannot and should not be owned or managed by government, by the rules and regulations of bureaucracy, hundreds of miles away, responding to a different constituency.

The measure of the success of this or any other program will be the extent to which it helps the ghetto to become a community—a functioning unit, its people acting together on matters of mutual concern, with the power and resources to affect the conditions of their own lives. Therefore the heart of the program, I believe, should be the creation of Community Development Corporations, which would carry out the work of construction, the hiring and training of workers, the provision of services, and encouragement of associated enterprises . . .

But a further and critical element in the structure, financial and otherwise, of these corporations should be the full and dominant participation by the residents of the community concerned . . .

At least for matters of immediate neighborhood concern, these Community Development Corporations might return us partway toward the ideals of community on a human scale, which is so easily lost in metropolis . . .

One purpose for which they must be an instrument, however, and one purpose which must be served by every aspect of the program I have proposed, or any other program, is to try to meet the increasing alienation of Negro youth . . . Among Negro youth we can sense, in their alienation, a frustration so terrible, an energy and determination so great, that it must find constructive outlet or result in unknowable danger for us all. This alienation will be reduced to reasonable proportions, in the end, only by bringing the Negro into his rightful place in this nation. But we must work to try and understand, to speak and touch across the gap, and not leave their voices of protest to echo unheard in the ghetto of our ignorance . . .

For all these programs, of course, there is a question of cost . . .

The financial question should be explored in this hearing, for it has the most direct and fundamental relevance to the problems of the city. Necessary as that exploration is, however, it should not be allowed to obscure the more fundamental question: Do the agencies of government have the will and determination and ability to form and carry out programs which do not fit on organization charts, which are tailored to no administrative convenience but the overriding need to get things done? If we lack this central ability, then vast new sums will not help us. The demonstration-cities proposal is a creative beginning, but it must be followed up by a demonstration of this critical ability to get things done, or the sums needed will not be forthcoming . . .

We do not only want to remedy the ills of the poor and oppressed—though that is a huge and necessary task—but to improve the quality of life for every citizen of the city, and in this way to advance and enrich American civilization itself.

The hearings that Kennedy's testimony helped to kick off featured appearances by many of the nation's big-city mayors. When New York City's John Lindsay presented (cavalierly, in Kennedy's eyes) a wish list for that city with no estimate of costs, Kennedy pressed him until an exasperated Lindsay pegged his request at $50 billion of federal aid. Kennedy's questions, and his dismissal of Lindsay's proposal as naive, were prompted by substantive disagreement: Kennedy felt every program should disclose its cost, and he thought Lindsay's proposal was overly reliant on federal funds and direction. Kennedy was also motivated by political calculation: Lindsay was widely rumored to be the likely Republican challenger to Kennedy's Senate reelection effort in 1970.

The hearings helped Kennedy conceptually, as testimony made clear that tax incentives could attract private business back to the inner cities; thereafter, such incentives were a standard component of Kennedy's programs. They also facilitated two legislative initiatives of great interest to Kennedy: the Model Cities Program and the Special Impact amendment to the Economic Opportunity Act (ultimately referred to as Title ID, establishing Community Development Corporations), which he helped enact that year, providing a funding source for the Bedford-Stuyvesant project, on which Kennedy placed so much hope.

Launching the Bedford-Stuyvesant Restoration Effort

BROOKLYN, NEW YORK

DECEMBER 10, 1966

During 1966, federal domestic spending continued to slow and President Johnson increasingly regarded each Kennedy policy initiative as a political power grab. Although Kennedy would by no means give up on a major federal role in rebuilding the country's cities, these constraints and his evolving views of the components of effective action led him to try to create a successful revitalization model away from Johnson's direct control. He had instructed aides to analyze initiatives from around the nation and to search for a demonstration locale in New York, in the aftermath of his January 1966 trilogy of speeches. Their attention quickly focused on the Bedford-Stuyvesant section of Brooklyn.

Bed-Sty, as the neighborhood is still called, encompasses nearly five square miles, and in the 1960s it was home to between a third and a half million people: 84 percent black, 12 percent Puerto Rican. It was larger than Fort Worth, Texas, or Toledo, Ohio, and was the nation's second-largest ghetto, surpassed only by Chicago's South Side. Bed-Sty had received almost no federal aid. The community did have comparative strengths over better-known Harlem: a superior housing stock;

a far higher percentage of home ownership; a bedrock of working residents; and the absence of the sclerotic, corrupt local power structure that had frustrated past efforts in Harlem. But visible signs of neglect greeted Kennedy as he walked through the area ten days after delivering his three January speeches. Many of the brown car hulks alternated with burned-out buildings, and the stench and sense of hopelessness seemed pervasive.

If Kennedy expected a hero's welcome when he met with community leaders after his walking tour, he was sorely mistaken. "Senator," one large man told him, somewhat hostilely, "we have been studied, examined, sympathized with, and planned for. What we need is action!"

Kennedy promised to return within the year with a plan to do just that. He arranged federal start-up funding by securing summer passage of his Special Impact Program amendment to 1966's principal poverty legislation, the Economic Opportunity Act. The initial federal commitment to Bed-Sty was $7 million for two years.

Simultaneously, he and his staff conducted detailed discussions with a wide range of residents. In 1964 testimony (while attorney general), Kennedy had said that planning "programs for the poor, not *with* them" was responsible for part of the "sense of helplessness and futility" that characterized ghetto life. In Bedford-Stuyvesant, he was determined to plan *with* the community. At the same time, Kennedy met with captains of industry, recognizing that private investment and active business involvement would be required for the neighborhood to thrive. To enhance the business community's comfort with the plan, he enlisted key Republican supporters, including his Senate colleague Jacob Javits and influential corporate leaders.

By the end of 1966, the pieces were in place. On December 10, Kennedy spoke to about one thousand people gathered in the auditorium of Public School 305, on Monroe Street in Bedford-Stuyvesant.

If men do not build," asks the poet, "how shall they live?"

That [is the] first question millions of men and women all over America ask themselves—ask us—every day; every day of idleness, every "day that follows day, with death the only goal."

That is the question, indeed, of life in the American city in years to come . . .

The plight of the cities—the physical decay and human despair that pervades them—is the great internal problem of the American nation, a challenge which must be met. The peculiar genius of America has been its ability, in the face of such challenges, to summon all our resources of mind and body, to focus these resources, and our attention and effort, in whatever amount is necessary to solve the deepest and most resistant problems. That is the commitment and the spirit required in our cities today.

And that is the spirit of this community, the spirit that is here today. Bedford-Stuyvesant, like other areas in the great cities all over America, has serious problems. This is a community in which thousands of heads of families, and uncounted numbers of young people, sit in idleness and despair; a community with the highest infant mortality rate in the city, one of the highest in the nation; in which hundreds of buildings we abandoned to decay, while thousands of families crowd into inadequate apartments. This is also a community long bypassed and neglected by government, receiving almost nothing out of the hundreds of millions of dollars the federal government gave to the city over two decades, unable to secure a single urban-renewal grant in ten years of trying.

But for all these difficulties, the spirit of Bedford-Stuyvesant has lived, the *community* has survived . . .

For the last eight months, I have had the privilege of seeing and working with [community leaders] at first hand.

Eight months ago, we found our views on the crisis before us to be in close correspondence. You through a manifesto of the Central Brooklyn Coordinating Council, and I in a series of speeches on the urban crisis, each proposed programs to meet this crisis in a comprehensive and coordinated effort, involving the resources and energies of government, of private industry, and of the community itself.

. . . We set our aim as a vital, expanding economy throughout the community—creating jobs in manufacturing and commerce and service industries.

On this basis of employment, we proposed the creation of new educational opportunities of many kinds . . . And we urged the reconstruction of social services, and their integration with the rebuilding effort . . .

Through the fabric of all program components, as I emphasized in

- cooperation with the private business community in self-sustaining economically viable enterprises;

- integration of programs for education, employment, and community development under a coordinated overall plan;
- and impetus and direction to be given in these efforts by the united strength of the community, working with private foundations, labor unions, and universities, in Community Development Corporations organized for this purpose . . .

But more important than any material component, you were determined that . . . these changes would happen, not by fiat from Washington, not from the offices of a president or a senator or a mayor, but from the work and effort of the Bedford-Stuyvesant community. You knew that what is given or granted can be taken away, that what is begged can be refused; but that what is earned is kept, that what is self-made is inalienable, that what you do for yourselves and for your children can never be taken away.

As a result of all of this—the fruit of eight months of planning and argument and exchange of views . . . I have the honor to announce:

First: The formation of the Bedford-Stuyvesant Renewal and Rehabilitation Corporation . . . with a distinguished board representing many elements of the community. This corporation will assume a major role in the physical, social, and economic development of the community.

Second: To work in closest partnership with the Renewal and Rehabilitation Corporation, there is being formed a Bedford-Stuyvesant Development and Services Corporation. This corporation will involve, and draw on the talents and energies and knowledge of, some of the foremost members of the American business community . . .

Major private foundations have committed support to the development of Bedford-Stuyvesant . . .

Many other organizations and individuals are contributing their energy and talent to Bedford-Stuyvesant . . .

Efforts have begun to secure means of financing the necessary development. Use of government funds is being developed with relevant federal and city agencies, including those concerned with housing and urban matters, manpower development, and education. And major efforts to attract private capital are also projected . . .

It is a beginning. Bedford-Stuyvesant is on its way. That way, as I will stress again and again, is not easy. It is complex and complicated and fraught with difficulty. Ahead of us are not weeks or months of work, no quick or easy triumphs—but long years of painful effort, with many setbacks, with constant temptations to relax, to give up, to stop trying . . .

To turn promise into performance, plan into reality . . . we must combine the best of community action with the best of the private enterprise system. Neither by itself is enough; but in their combination lies our hope for the future . . .

If there is to be any action, any true progress in a community, that community itself . . . must be prepared to take full and final responsibility for what happens—for the success or failure of any program . . .

But for all that community action can do, for all the talent and energy it may liberate, still it is not enough. For it does not give the power to act: not just to petition but itself to act to improve the lives of people . . . People may ask or even protest for better community services or quality goods in the stores, but concessions wrung from an unwilling bureaucrat or absentee owner will never equal, in quality or permanence, the quality of service that can be created or bought by the united resources of a self-reliant community with the resources to act for itself.

The power to act is the power to command resources, of money and mind and skill: to build the housing, create the social and educational services, and buy the goods which this community wants for action, and needs and deserves. The regeneration of the Bedford-Stuyvesant community must rest, therefore, not only on community action but also on the acquisition and investment of substantial resources in this area. That is the importance and function of the Renewal and Rehabilitation Corporation, and particularly of the Development and Services Corporation: to stimulate and facilitate the investment of resources from the private business community, in conjunction with foundation and government support, in Bedford-Stuyvesant. Such investment will have multiple benefits. It will help to build the housing and services which will make this a better place to live. And, by providing jobs for area residents, it will create a sound economic base—a foundation of self-government and dignity—for the entire community . . . not to ask others to act.

Private enterprise will invest in Bedford-Stuyvesant only if it can be assured that this community, acting as a unit, is prepared to deal with private capital on the businesslike basis; that it will and can, acting through a Renewal and Rehabilitation Corporation and the Development and Services Corporations, offer:

- a place to locate;
- people willing to work and learn;
- programs to train workers;
- and all other services necessary to operate.

Government and foundations will only provide the needed incentive and support money if they know that programs are soundly conceived and operated; that important positions are assigned on no grounds other than merit; that there is no room here for political dealing, or for jobs to be regarded as anything but.

And if this is true for outside investment, it is even more true for the people of Bedford-Stuyvesant. The people of this area will be asked to make sacrifices—of time and convenience and effort. More importantly, Bedford-Stuyvesant wants to command its own destiny, and this will require direct investment by its own people. But if this is to take place, then the people must have faith in the programs and their leadership . . .

[There] is the need for cooperation between the community and all those—the businessmen, and the public officials, and the experts—who are joined with you in this effort. This community, in the last analysis, must do the job itself . . .

But at the same time, we will all have to listen and consider most carefully the advice, and the recommendations, and sometimes the absolute requirements, of others. If a government program requires a certain standard of operation, that standard must be maintained. If a businessman requires a certain kind of training program to help him offer jobs to people here, then that kind of training program must be devised. If banks require a certain kind of feature in a financing arrangement before they will make loans for housing, those arrangements must be satisfactorily made. If the city needs to coordinate efforts in Bedford-Stuyvesant with efforts elsewhere, then cooperation must be given . . .

What remains is the heart of the matter; and fulfillment will be the hardest part of the task. There will be times when progress seems ephemeral and fleeting, times of great disappointment and discouragement. Always there will be work—ceaseless, untiring effort, by none as much as the people in this room.

For this is a task of unparalleled difficulty. This is not just a question of making Bedford-Stuyvesant "as good as" someplace else. We are striking out in new directions, on new courses, sometimes perhaps without map or compass to guide us. We are going to try, as few have tried before, not just to have programs like others have, but to create new kinds of systems for education and health and employment and housing. We here are going to see, in fact, whether the city and its people, with the cooperation of government and private business and foundations, can meet the challenges of urban life in the last third of the twentieth century.

But if the dangers are great, and the challenges are great, so are the possibilities of greatness. In the last month, we have come to know one another well; and I believe that we can succeed, that we can fulfill the commitment, and thereby help others to do so.

And so let us go forward . . .

Kennedy remained actively involved in the Bed-Sty effort for the remainder of his life. "It was his work, his vision, energy, enthusiasm, and intelligence," remembered John Doar (recruited out of the Justice Department in 1967 to run the business sector's involvement in the community), "that kept it going."

The Bedford-Stuyvesant Restoration Corporation has made a positive difference for Bed-Sty over the intervening years, constructing or rehabilitating thousands of housing units, creating tens of thousands of jobs, and reaching nearly one hundred thousand residents annually with its programs. The organization also survived the federal government defunding Community Development Corporations in the 1980s, in part by securing financial support from the corporate community. Successful city models today often trace their roots to Bed-Sty and other early CDCs, involving community and the private sector in wide-ranging programs.

A Holiday Reflection for White America: Citizens Union

NEW YORK, NEW YORK

DECEMBER 14, 1967

While Kennedy was seeing the Bedford-Stuyvesant Restoration project begin to bear fruit in 1967, Lyndon Johnson was surrendering in the War on Poverty. The slogan that identified his domestic dreams, the "Great Society," no longer was used in presidential speeches. Innovations to end poverty and promote racial justice gave way, wrote Daniel Patrick Moynihan (later to become a Democratic senator from New York) in 1968, to "disquisitions on Safe Streets and Crime Control Acts, and other euphemisms for the forcible repression of black violence."

And violence defined the summer of 1967. Beginning in the South in May, rioting spread north: In six July days of rage in Newark, New Jersey, twenty-six people died. Less than a week later, violence broke out in New York's Spanish Harlem. Then, on July 23, 1967, Detroit exploded and forty-three people died in four days. Johnson sent armed helicopters, tanks, forty-seven thousand paratroopers, and three thousand National Guardsmen to the city. The president's explanation of his decision, delivered on national television, made no mention of any underlying causes of the riots. Kennedy too continued to insist on the rule of law. Speaking in San Francisco ten days after the Detroit riot began, he said: "We must make it unequivocally clear by word and deed that this wanton killing and burning cannot and will not be tolerated."

But Kennedy was more certain than ever that increased law enforcement was no answer. When his press secretary, Frank Mankiewicz, in despair over the president's inactivity and insensitivity, asked Kennedy what he would do if he were in Johnson's place, Kennedy said the first step after stopping the violence would be to cajole the television networks into showing the white majority what living conditions were like in the riot-ravaged areas:

> *Let them show the sound, the feel, the hopelessness, and what it's like to think you'll never get out. Show a black teenager, told by some radio jingle to stay in school, looking at his older brother, who stayed in school, and who's out of a job. Show the Mafia pushing narcotics; put a* Candid Camera *team in a ghetto school and watch what a rotten system of education it really is. Film a mother staying up all night to keep the rats from her baby . . . Then I'd ask people to watch it*

and experience what it means to live in the most affluent society in history—without hope.

Then, Kennedy said, he'd bring together business and community leaders and cobble together plans in each city the way he had in Bed-Sty.

But Kennedy wasn't president, and Lyndon Johnson was squandering the bully pulpit on the war in Vietnam. In *To Seek a Newer World,* Kennedy noted that "through the eyes of the white majority, of a man of decent impulse and moral purpose, the Negro world is one of steady and continuous progress . . . But if we look through the eyes of the young slum dweller . . . there is a different view, and the world is a hopeless place indeed."

Determined to show white America this other world, Kennedy delivered variations of the same speech to audiences in San Francisco in August, and, as follows, at New York's Citizens Union (a public interest watchdog organization founded in 1897) on December 14, 1967, in the middle of the holiday season.

Long ago it was written that "to everything there is a season, and a time to every purpose under the heaven."

Now we approach the season which, in every year, we mark our higher purposes and our common humanity.

In this year, this season is also a time to pause and a time to reflect, on where we have been and where we are going—halfway between the terrible summer of 1967 and what may well be a most difficult time in the summer of 1968 . . .

The weather is cool now, and the violence that breaks out, breaks out only occasionally.

But the embers of disaffection and distrust, the sparks of frustration and discontent—these still burn and these still crackle, in the ashes of Detroit and Newark, Los Angeles and New York, and in every city across this nation.

We know that this winter is worse than last winter, and that next summer may be worse than last summer; a grotesque spiral of greater violence and ever-greater vengeance, threatening the well-being and liberties of every American citizen.

We know that this will happen—unless we act, now, to heal the division and cure the sores in which violence festers.

Since the events of last summer this has been plain to all with the eyes to see.

We have seen—but we have not acted.

Yet only action—forceful, direct, sweeping action—can meet our problems . . .

Look through the eyes of the young slum dweller—the Negro, the Puerto Rican, the Mexican American—at the dark and hopeless world he sees . . .

Kennedy then described the horrifying statistics of broken families, dying babies, failed schools, and demoralizing unemployment that characterized America's cities. He continued:

On his television set, the young man can still watch the multiplying marvels of white America; the commercials still tell him life is impossible without the latest products of our consumer society.

All this goes on.

But he still cannot buy them.

How overwhelming must be the frustration of this young man—this young American—who, desperately wanting to believe and half believing, finds himself still locked in the slums, his education second rate, unable to get a job, confronted by the open prejudice and subtle hostilities of a white world, and powerless to change his condition or even have an effect on his future.

Others still tell him to work his way up as other minorities have done; and so he must.

For he knows, and we know, that only by his own efforts and his own labor will he come to full equality.

But how is he to work?

The jobs have fled to the suburbs, or have been replaced by machines, or have moved beyond the reach of those with limited education and skills . . .

The fact is, if we want to change these conditions—those of us here in this room, those of us who are in the establishment, whether it be business, or labor, or government—we must act.

The fact is that we can act.

And the fact is also that we are not acting . . .

Kennedy next described his program for urban revitalization, most of which he had developed in 1966.

We must turn the power and resources of our private enterprise system to the underdeveloped nation within our midst.

This should be done by bringing into the ghettos themselves productive and profitable private industry—creating dignified jobs, not welfare handouts, for the men and youth who now languish in idleness. To do this, private enterprise will require incentives—credits, accelerated depreciation, and extra deductions—as effective and comprehensive as those we now offer for the production of oil or the building of grain-storage facilities or the supersonic transport . . .

But what is far beyond doubt is that the resources and abilities of private industry, and not just the federal treasury and bureaucracy, must be engaged in this great task . . .

There is, after all—for all of us—no alternative.

History has placed us all, black and white, rich and poor, within a common border and under a common law.

All of us, from the wealthiest to the young children that I have seen in this country, in this year, bloated by starvation—we all share one precious possession, and that is the name *American.*

It is not easy to know what that means.

But in part to be an American means to have been an outcast and a stranger, to have come to the exiles' country, and to know that he who denies the outcast and stranger still amongst us, he also denies America.

That is something for us to consider, in this Christmas of 1967.

Rural Poverty, Exploitation, and Hunger in America

Rural Poverty Hearings

WASHINGTON, D.C.; MISSISSIPPI; AND KENTUCKY

MARCH, APRIL, AND JULY 1967; FEBRUARY 1968

Kennedy, as a member of the Migratory Labor Subcommittee of the Senate Labor Committee, heard testimony in his first year in the Senate about the wretched living and working conditions of the laborers who followed the harvests. Late in 1965, friends in organized labor asked Kennedy to take the subcommittee to the San Joaquin Valley of California, where Cesar Chavez, trained by legendary community organizer Saul Alinsky, was unionizing grape pickers and urging a national boycott of nonunion grapes. Kennedy initially tried to avoid taking on another issue; he was resistant, preoccupied by Vietnam and the problems of the cities. But after more prodding from labor, he reluctantly joined his subcommittee for the second day of hearings in a high school auditorium in Delano, California, in March 1966.

What he saw and heard galvanized him. Chavez remembered the day as stiflingly hot and the auditorium as jam-packed. "There were workers inside and outside and crawling out the windows and in the doors." Seats had to be cleared for the growers, who were, in Chavez's understatement, "very hostile." Kennedy listened to testimony from farm workers and from growers (who argued that the organizers were Communists or their dupes). Chavez later paraphrased Kennedy's counsel to the growers: "It's not in my mind that you're going to win. I would suggest, of course, in terms of this committee, that you sit down and negotiate before the thing gets worse, because the workers are going to win."

Kennedy's endorsement of their struggle was politically and psychologically important to Chavez's movement. Union leader Dolores Huerta remembered that "Robert didn't come to us and tell us what was good for us. He came and asked two questions . . . 'What do you want?' and 'How can I help?' That's why we loved him." He stayed in contact with the movement over the next two years, raising donations for them; working for minimum wage and collective bargaining legislation that would include itinerant workers such as grape pickers; and fighting to have the Immigration and Naturalization Service stop using threats of deportation to break the union. Kennedy had great personal respect for Chavez, a few years his junior, particularly because of Chavez's shrewd organizing talents and his commitment to nonviolence in the struggle for increased economic and political power for migrant workers.

The exploitation and numbing poverty Kennedy saw afflicting California's farm workers also characterized much of the South. In 1967, the legislative authorization for much of the War on Poverty was due to expire, requiring reenactment. On March 15, 1967, the Labor Subcommittee on Employment, Manpower, and Poverty began hearings intended to generate national interest and support for further funding and to dramatize the extent of hunger in the country. One of the first to testify at the Capitol was Marian Wright, a twenty-seven-year-old southern native and Yale Law School graduate, who staffed the NAACP Legal Defense and Education Fund's Mississippi office.[6] She was one of only five black lawyers (and one of two black law school graduates) in the state, and the only one with a full-time specialty in civil rights.

Wright and other witnesses spoke of the worst poverty seen since the Great Depression in the thirties, as agricultural mechanization eliminated the handpicking and planting jobs on which thousands of southern blacks had depended for generations. By 1967, many had no income whatsoever. Unable to afford the monthly payments required for the new food stamps program, many were malnourished, and some were dying from hunger.

Kennedy and Subcommittee Chair Joseph Clark began formal hearings in Mississippi on April 10, 1967. Marian Wright's testimony set the context: "After two civil rights bills and the third year of the poverty bill, the . . . Negro in Mississippi is poorer than he was, he has less housing, he is badly educated; he is almost in despair."

In the evening before the hearings, and throughout the full day of testimony, Kennedy heard tale after tale of appalling conditions and explanations from officials who ranged from overburdened to unconcerned. The state provided welfare assistance only to broken homes, and the aid was nine dollars per child, per month, up to a maximum total of ninety dollars per month. Under new rules, participation in federal food programs had been cut in half in some counties. But Mississippi's Governor Paul Johnson, when asked that year about hunger among blacks in his state, scoffed, "All the Negroes I've seen around here are so fat they shine!"

After the hearing, Kennedy told Charles Evers (brother of the slain Medgar and a key contributor to Kennedy's black support in the 1964 Senate campaign), "I want to see it." Led by local civil rights leader Amzie Moore, the small group began in Cleveland, Mississippi, and ventured deep into the Delta, into (in Evers's words) "one of the worst places [I have] ever seen." They went from shack to shack, listening to men long out of work, without skills or prospects, and parents trying to keep their babies alive on rice, biscuits, and gravy left over from old surplus handouts. Kennedy told traveling companion Edelman that although he had witnessed serious deprivation in West Virginia, he was viewing the most horrendous conditions he had seen in the United States, equal to the worst squalor he had seen in the Third World.

Evers remembered one shack with

> *no ceiling hardly. The floor had holes in it, and a bed . . . black as my arm, propped up with some kind of bricks to keep it from falling. The odor was so bad you could hardly keep the nausea down . . . This lady came out with hardly any clothes on, and we . . . told her who [Kennedy]*

> *was. She just put her arms out and said "Thank God" and then she just held his hand.*

Des Moines Register writer Nick Kotz, accompanying the group, watched Kennedy, who had seen "a child sitting on the floor of a tiny back room. Barely two years old, wearing only a filthy undershirt, she sat rubbing several grains of rice round and round on the floor. The senator knelt beside her.

> *"Hello . . . Hi . . . Hi, baby . . ." he murmured, touching her cheeks and her hair as he would his own child's. As he sat on the dirty floor, he placed his hand gently on the child's swollen stomach. But the little girl sat as if in a trance . . . For five minutes he tried: talking, caressing, tickling, poking—demanding that the child respond. The baby never looked up.*[7]

Evers remembered that "tears were running down [Kennedy's] cheeks, and he just sat there and held the little child. Roaches and rats were all over the floor . . . 'How can a country like this allow it?' [Kennedy asked him]. 'Maybe they just don't know.'"

Because of Kennedy's prominence, the tour had been accompanied by a television crew, and that night many

Americans got their first views of the rural hardship Kennedy had just witnessed.[8]

Kennedy returned to Washington and met immediately with Agriculture Secretary Orville Freeman. The secretary doubted conditions were desperate and pointed to existing federal programs. Freeman feared Kennedy's political wrath, but he was even more concerned about the reactions of a group of southerners in Congress who controlled his budget and who regarded Kennedy and Clark as troublemakers.

Kennedy tried to involve the White House. Armed with the report of Freeman's team of nutritionists, who found that hunger was "uniformly" worse for the poor in Mississippi when compared to twelve years earlier, the full subcommittee formally petitioned the president to declare an emergency. Johnson never replied.

Trying to keep up the pressure on Freeman to release emergency funds for food, Kennedy and Clark called follow-up hearings for July 11 and 12, 1967. The eloquence of witnesses from a team of doctors who had undertaken a detailed examination of hungry and diseased children throughout Mississippi temporarily defused the attacks of the state's two senators. One of them, John Stennis, proposed $10 million in emergency hunger and medical aid, hoping to contain any large-scale federal intervention. Kennedy and Clark quickly shepherded the idea through the Senate, increasing the allotment, but the powerful chairman of the House Agriculture Committee, Texan William Poage, killed the bill when it crossed the Capitol. Finally, in November, Kennedy helped to tack the Stennis bill on to the reauthorization of the entire poverty program, and the amended measure passed. Even then, the administration stalled its implementation, and only in April 1968 did the "emergency" aid finally begin to be dispensed.

Migrant Workers

DELANO, CALIFORNIA, AND MONROE COUNTY, NEW YORK

MARCH 16, 1966; SEPTEMBER 8, 1967; MARCH 10, 1968

In September 1967 Kennedy returned to the problems of migrant workers, this time in his own state. (As Kennedy was fond of pointing out to audiences in the farm belt during the 1968 presidential campaign, upstate New York actually led the nation in the production of some agricultural commodities.) Accompanied by fellow New York Senator Jacob Javits, Kennedy traveled to a work camp outside of Rochester owned by Jay DeBadt. DeBadt, armed with a shotgun, kept the entourage of twenty at bay. Kennedy sent Edelman around to the back to secure an invitation from the migrant workers encamped there; when Edelman returned, Kennedy walked past DeBadt with a terse "I was invited on this property." Most of the others remained at the entrance, in a tense confrontation with the owner.

Kennedy, accompanied by writer Jack Newfield, walked into a derelict bus that served as living quarters for three families. It was filthy and the odor was overpowering. Inside, they found children covered "with unhealed scabs and flies."

"Bobby's hand began to shake from . . . rage," Newfield recalled. "It was like watching a man going through a religious conversion. Then an old, bent Negro woman wandered into the bus, and Bobby, on instinct, reached out to touch the back of her neck. He asked her how much she earned, and she told him one dollar an hour for picking celery."

Returning to the others, Kennedy was confronted by an enraged DeBadt: "You're just a do-gooder. You're here just trying to grab a few headlines. Get off my property."

Struggling to control himself, Kennedy hoarsely replied, "You're something out of the nineteenth century. I wouldn't let an animal live in those buses."

The following day, Kennedy wrote to New York Governor Nelson Rockefeller, demanding an investigation of health conditions in the camps. Now under scrutiny, the authorities moved in. Kennedy also pressed labor leaders to organize the migrant workers, as Cesar Chavez was doing in the California grape fields.

To maintain the public spotlight on the problems of rural poverty, Kennedy convinced Chairman Clark to let him head another round of hearings. Originally he had intended to go to South Carolina but, persuaded by his Democratic colleague Senator Fritz Hollings that the notoriety would harm Hollings's reelection bid, Kennedy instead visited Kentucky for sessions on February 13 and 14, 1968. He summarized them, and his various experiences in Mississippi, Rochester, and Delano, in a speech he intended to deliver later in the month at Kansas State University.

Why is the most outwardly successful nation in the world so troubled in spirit, so rent by division, so seemingly purposeless and adrift? . . .

First, in this richest of nations, we have allowed the perpetuation of poverty and degradation which can only be described as outrageous. In the ghettos of the great cities are hundreds of thousands of men without work . . .

But there are others from whom we avert our sight. Some of them are in the hills and hollows of [the] Appalachians. That is proud land and these are proud men, who have rallied to the nation's flag at every hour of danger. But the deep mines are closing, and the jobs have gone, leaving men without work, many of them crippled by the accidents and disease that lurk "down in the mines," their land a ruin of strip mines and stinking creeks.

Their children are ravaged by worms and intestinal parasites. They eat bread and gravy and sometimes beans; and as one of them says, when another child is born "we just add a little water to the gravy."

They work, when they can, on governmental projects: projects so meager that there are often no tools to work with—nothing to help them build roads through the mountains but a forked sapling branch like that the first men may have used to till the soil five hundred centuries ago.

And there are others: on the back roads of Mississippi, where thousands of children slowly starve their lives away, their minds damaged beyond repair by the age of four or five; in the camps of the migrant workers, a half million nomads virtually unprotected by collective bargaining or Social Security, minimum wage, or workmen's compensation, exposed to the caprice of fate and the cruelty of their fellow man alike; and on Indian reservations where the unemployment rate is 80 percent, and where suicide is not a philosopher's question but the leading cause of death among young people.

Only a minority are poor. But poverty affects all of us . . . the facts of poverty and injustice penetrate to every corner, every suburb and farm in the nation. Their existence is the message of every evening news broadcast, crippling our satisfaction in our ownership of one, or two, or three of America's seventy million television sets . . .

Our ideal of America is a nation in which justice is done; and therefore, the continued existence of injustice—of unnecessary, inexcusable poverty in this most favored of nations—this knowledge erodes our ideal of America, our basic sense of who and what we are. It is, in the deepest sense of the word, demoralizing—to all of us.

The speech was never delivered. By late February 1968, Kennedy was readying a challenge to Lyndon Johnson for the presidency, and the trip to Kansas was postponed (it became one of the first presidential campaign stops, and the draft language served as the source of the early portions of Kennedy's March 18 speech at the University of Kansas). Another journey to help the impoverished remained, for in the middle of February, Cesar Chavez began a fast to inspire the farm workers and recommit the struggle to the tenets of nonviolence.

Edelman learned from friends in Delano that the fast was going unnoticed by the national press. Some time after his second week without food or liquids, Chavez's condition increasingly concerned his doctors, who relayed to Kennedy a message that he alone might be able to urge the union leader to eat. Kennedy began to monitor the situation daily, and when Chavez agreed to end the fast (after twenty-five days) on Sunday, March 10, 1968, Kennedy flew west as the only public official invited to join in the Mass of Thanksgiving. His advisers reminded him of the red-baiting that followed his 1966 visit and worried that the trip might cost Kennedy votes in the crucial California primary. "I know," Kennedy told them, "but I like Cesar."

Chavez greeted Kennedy from his bed. The circumstances made both of them awkward ("What do you say to a man who's on a fast?" Kennedy had asked UAW leader Paul Schrade on the way to Delano. "Just say, 'Hello, Cesar,'" Schrade advised). Kennedy was deeply moved and, accompanying the weakened Chavez to Delano Memorial Park, was mobbed by some six thousand emotional farm workers who crowded around a flatbed truck holding a makeshift altar. Having lost thirty-five pounds, Chavez was too weak to address the gathering. After he had broken his fast by sharing a bit of *semita* (round, anise-flavored Mexican-style bread) with Kennedy, and Father Mark Day's formal mass had ended, Kennedy addressed the gathering, his hands scratched and bleeding from the physicality of the crowd's welcome. He stumbled through a prepared greeting, phonetically written out in Spanish, and turned to Chavez: "I'm murdering the language, Cesar, is that right?" "Yes," the smiling union leader agreed. Kennedy then continued, in English:

This is a historic occasion. We have come here out of respect for one of the heroic figures of our time—Cesar Chavez. But I also come here to congratulate all of you, you who are locked with Cesar in the struggle for justice for the farm worker, and the struggle for justice for the Spanish-speaking American. I was here two years ago, almost to the day. Two years ago your union had not yet won a major victory. Now elections have been held on ranch after ranch and the workers have spoken. They have spoken, and they have said, "We want a union."

You are the first—not the first farm workers to organize, but the first to fight and triumph over all the odds, without proper protection from federal law.

You have won historic victories . . . You have won them with your courage and perseverance. You stood for the right; you would not be moved.

And you will not be moved again.

The world must know, from this time forward, that the migrant farm worker, the Mexican American, is coming into his own rights. You are winning a special kind of citizenship. No one is doing it for you—you are winning it yourselves—and therefore no one can ever take it away.

And when your children and grandchildren take their place in America—going to high school, and college, and taking good jobs at good pay—when you look at them, you will say, "I did this. I was there, at the point of difficulty and danger." And though you may be old and bent from many years of labor, no man will stand taller than you when you say, "I marched with Cesar."

But the struggle is far from over. And now, as you are at midpoint in your most difficult organizing effort, there are suddenly those who question the principle that underlies everything you have done so far: the principle of nonviolence. There are those who think violence is some shortcut to victory.

You are serving under a leader of this union who is committed to the principle of nonviolence. And that's so important for any success that you are going to have in the future—that we respect the principle of nonviolence. If there was anything that the unions learned and others learned during the 1930s, it was the fact that violence brought no answer. And if there is anything that we've learned during the 1960s, all of us who are here, it is that violence is not the answer to our problems.

And let no one say that violence is the courageous way, that violence is the short route, that violence is the easy route. Because violence will bring no answer: It will bring no answer to your union; it will bring no answer to your people; it will bring no answer to us here in the United States, as a people.

And that's why I come here today to honor Cesar Chavez, for what he's done and what he stands for. How desperately you, and the people of this country—of our country—need him today. I come here today to honor you for the long and patient commitment you have made to this struggle for justice. And I come here to say that we will fight together to achieve for you the aspirations of every American: decent wages, decent housing, decent schooling, a chance for yourselves and your children. You stand for justice, and I am proud to stand with you.

Viva La Causa.

On the way to Delano, Kennedy had told Ed Guthman (then–national editor for the *Los Angeles Times)* that he would announce his bid for the presidency after the

New Hampshire primary, only two days away. After the Mass, he also told Chavez. The farm workers would stand with him in the campaign ahead. As Chavez remembered, the intensity of their feelings for Kennedy "was a phenomenon that can't be explained . . . It's that line that you very seldom cross—I've never seen a politician cross that line and I don't think that I'll ever live to see another public figure [do it]."

Filling Out a Domestic Agenda

By the time he broke bread with Chavez and prepared for his struggle on the presidential campaign trail, Kennedy had confounded many of his Senate colleagues during his more than three years in the Capitol. When he took office in 1965, many assumed that his reputed aggressive individualism would render him ineffective in such a ritualistically collegial body, and that his presumed ambition for the presidency would make the Senate a way station, not a place for substantive accomplishment. Kennedy had proven them wrong.

His passion and persistence regarding emergency food assistance for Mississippi's poor showed that he could channel his impatience, marshal the media, work with his colleagues, outmaneuver the administration, and win a small but signal victory for people in dire need. As a war where most of the battles had to be won by staff, it also demonstrated Kennedy's ability to field an unparalleled cadre of aides—a characteristic that had marked his team of investigators on the Rackets Committee, his campaign structure in 1960, and his staff at the Justice Department.

If Kennedy's activism on behalf of the nation's poor, hungry, and disenfranchised marked a recurring chorus in his years in the Senate, there were myriad other melodies. He authored and directed the passage of important amendments to legislation concerning education (establishing standards and accountability for federally mandated programs), housing, narcotics, Medicare, health services, voting rights, economic development, and cigarette labeling and advertising, among others. In addition, Kennedy presented a detailed campaign-finance reform package in committee testimony, and he introduced a sophisticated bill to overhaul Social Security.

To demonstrate this breadth, four speeches are excerpted. The first two, delivered in 1966 in upstate New York and suburban Minnesota, respectively, indicate the extent to which Kennedy was committed to transcending liberal dogma, by demanding that well-intentioned but ineffective government programs be redirected or scrapped, and by advocating a renewed commitment to individual effort and responsibility in the service of the

Redirecting Government, Solving Problems

UTICA, NEW YORK

FEBRUARY 7, 1966

Kennedy frequently called for greater federal involvement and often for increased spending, but he was an early and consistent critic of equating program-passing with problem-solving. He enunciated his views before the New York State Society of Newspaper Editors on February 7, 1966.

We stand at a new crossroads to an uncertain future. The way ahead is not charted. We know only that it will be full of difficulty and danger.

It may seem strange to talk now of new crossroads and new turnings. We are, after all, in the midst of the longest period of sustained expansion in our history. Our power and wealth are greater than ever before. More of our children go on to college, and even today's high school students learn as much as some collegians of yesterday.

Every day, some voyager to the frontiers of science returns with tidings of new possibilities, new vistas, new opportunities for ourselves and our children.

And following the election of 1964, the federal government fulfilled old dreams by the dozens: Medicare, aid to education, voting rights, immigration reform.

The inheritance of the New Deal is fulfilled. There is not a problem for which there is not a program. There is not a problem for which money is not being spent. There is not a problem or a program on which dozens or hundreds or thousands of bureaucrats are not earnestly at work.

But does this represent a solution to our problems? Manifestly it does not. We have spent ever-increasing amounts on our schools. Yet far too many children still graduate totally unequipped to contribute to themselves, their families, or the communities in which they live . . .

We have spent unprecedented sums on buildings of all kinds. Yet our communities seem less beautiful and sensible every year.

We have spent billions for agricultural price supports. Yet the rural economy continues to decline, and more and more people leave the farms for life in the urban centers.

We have spent billions on armaments and on foreign aid. Yet the world is still unsafe, and our position more precarious and painful as time goes by.

Why is this so?

The answer is one we have always known, though sometimes forgotten.

Money by itself is no answer. Programs which are misdirected accomplish nothing. To work at something merely because it does no devious harm is not enough.

There are things more important than spending.

Their names are imagination, courage, and determination.

And for those of us who speak to the public—to our fellow citizens—there is an especial fourth requirement: candor.

A few examples will illustrate.

One is welfare. Opponents of welfare have always said that welfare is degrading, both to the giver and the recipient. They have said that it destroys self-respect, that it lowers incentive, that it is contrary to American ideals.

Most of us deprecated and disregarded these criticisms. People were in need; obviously, we felt, to help people in trouble was the right thing to do.

But in our urge to help, we also disregarded elementary fact. For the criticisms of welfare do have a center of truth, and they are confirmed by the evidence.

Recent studies have shown, for example, that higher welfare payments often encourage students to drop out of school, that they encourage families to disintegrate, and that they often lead to lifelong dependency. Cecil Moore, head of the Philadelphia NAACP, once said that welfare was the worst thing that could have happened to the Negro. Even for such an extreme position, there is factual support.

Because most of us were committed to doing something we thought was good, we ignored the criticisms. But we also therefore ignored the real need—which was, and remains, decent, dignified jobs for all . . .

[Another] example: We have always operated our schools on the theory that the school system itself was right; if a child failed, it was the child who was at fault. So we put labels on children who failed. We called them culturally disadvantaged, or retarded, or perhaps lazy or stupid.

But the results of this policy are that one quarter to one third of our young men cannot meet even the minimal mental qualifications for the armed forces; that over half the graduates of many of our high schools are unequipped for even the most rudimentary jobs; that hundreds of thousands of children waste away in institutions.

We can no longer afford this waste. If our present educational methods cannot do better, then they must be changed to fit the students, just as doctors change a treatment which fails to cure a sick patient. We must now regard a student's failure as the school's failure, and as our failure—and hold ourselves responsible for our children's shortcomings.

. . . There is no question which does not require the same new thinking, the same willingness to dare . . .

We have committed our surplus food to feed the starving abroad, and we have offered to help in curbing population growth; but will we act on the scale necessary to prevent the mass starvation which our present level of effort cannot forestall?

We have abandoned isolationism, and made commitments in every corner of the world; but will we be equally ready to abandon the status quo, and associate ourselves with the rising forces of revolution—in Latin America, in Africa, in Asia? . . .

Lincoln said it best: "We must think anew and act anew. We must disenthrall ourselves."

To say it, however, is not to do it. It is not easy, in the middle of one's life or political career, to say that the old horizons are too limited, that our education must begin again, that new visions must replace the old if our vitality is to remain and be renewed . . .

[Yet] surely we must be determined to do this.

On Rebuilding a Sense of Community

WORTHINGTON, MINNESOTA

SEPTEMBER 17, 1966

Kennedy's call for reflection and readjustment stood in stark contrast to prevailing ideologies of both the political right (who saw no useful role for government outside of guaranteeing public safety) and the left (who were coming to regard large-scale federal intervention as the preferred remedy for any domestic ill). Kennedy also stood outside the mainstream's affection for modernity: the fascination with the new, the celebration of the big. On September 17, 1966, he spoke at a community college dedication in tiny Worthington, Minnesota.

Even as the drive toward bigness [and] concentration . . . has reached heights never before dreamt of in the past, we have come suddenly to realize how heavy a price we have paid: in overcrowding and pollution of the atmosphere, and impersonality; in growth of organizations, particularly government, so large and powerful that individual effort and importance seem lost; and in loss of the values of nature and community and local diversity that found their nurture in the smaller towns and rural areas of America. And we can see . . . that the price has been too high. Bigness, loss of community, organizations and society grown far past the human scale—these are the besetting sins of the twentieth century, which threaten to paralyze our very capacity to act, or our ability to preserve the traditions and values of our past in a time of swirling, constant change.

To these central dangers . . . we can trace a hundred others [in] the signs around us that all is not well in the republic: spreading violence, unconcern for others, too many seeking escape in noninvolvement or in drugs, debate become acerbic and bad tempered, and overall a sense that no one is listening.

Therefore, the time has come . . . when we must actively fight bigness and overconcentration, and seek instead to bring the engines of government, of technology, of the economy, fully under the control of our citizens, to recapture and reinforce the values of a more human time and place . . . It is not more bigness that should be our goal. We must attempt, rather, to bring people back to . . . the warmth of community, to the worth of individual effort and responsibility . . . and of individuals working together as a community, to better their lives and their children's future. It is the lesson that government can follow the leadership of private citizens: that men who are citizens in the full sense of the word need not belong to the government in order to benefit their community. And it is the lesson that if this country is to move ahead . . . it will not be by making everything bigger, not by piling all our people further on top of one another in huge cities, not by reducing the citizen to the role of passive consumer and recipient of the official vision, the official product.

Air Pollution Control Conference

NEW YORK, NEW YORK

JANUARY 4, 1967

As Kennedy noted in Minnesota, one of modernity's worst byproducts was pollution. On January 4, 1967, before the issue captured the widespread attention of either the public or politicians, Kennedy addressed one of the first major conferences to examine the consequences of fouling the environment: the New York-New Jersey Metropolitan Area Air Pollution Control Conference.

On a trip to Latin America last year, I saw people in Recife, in the poorest part of Brazil, who ate crabs which lived off the garbage that the people themselves threw in the shallow water near their shabby homes. And whenever I tell this story to Americans, the reaction is: How sad, how terrible, that such poverty, such underdevelopment, should exist in the world.

But we New Yorkers are in a poor position from which to extend pity. For every year, the average New Yorker—old and young, rich and poor, athlete or infirm recluse—breathes in 750 pounds of his own wastes . . .

And because there are so many of us, crowding into this tiny fraction of the United States, a great pall of filthy air blankets the entire metropolitan area, and we all must breathe the same air into which we carelessly spill our refuse . . .

But we should not—we cannot—wait for technology to make clean air entirely painless, to be achieved without effort, like a genie waving a magic wand. We will never get anywhere unless we begin now to apply what we do know . . .

We should, I believe, beware of the pitfalls described by Taine:[9] "Imagine a man who sets out on a voyage equipped with a pair of spectacles that magnify things to an extraordinary degree. A hair on his hand, a spot on the tablecloth, the shifting fold of a coat, all will attract his attention; at this rate, he will not go far, he will spend his day taking six steps and will never get out of his room."

We have to get out of the room.

Firearms Legislation Testimony

WASHINGTON, D.C.

JULY 11, 1967

Paralysis in the face of obvious need and present danger was not limited to inattention to pollution. Six months later, Robert Kennedy testified in favor of firearms legislation before the Subcommittee on Juvenile Delinquency of the Senate Judiciary Committee.

I am grateful for the opportunity to testify on a matter of deep national interest. Regulation of the sale of firearms is, in my judgment, essential for the safety and welfare of the American people.

Every year, thousands of Americans are killed by firearms . . . The great majority of these deaths and crimes would not have occurred if firearms had not been readily available. For the evidence is clear that the availability of firearms is itself a major factor in these deaths . . .

Basically, this bill would subject deadly weapons to a lesser control than we have always imposed on automobiles, liquor, or prescription drugs. The use and sale of these things are carefully regulated by federal, state, and local government. The same should be true of firearms . . .

Nevertheless, the nation, Congress, and sportsmen have been subjected to a massive publicity campaign against this bill. This campaign has distorted the facts of the bill and misled thousands of our citizens. Those responsible for this campaign place their own minimal inconvenience above the lives of the many thousands of Americans who die each year as the victims of unrestricted traffic in firearms. The campaign is doing the nation a great disservice.

And in recent weeks, the campaign has taken a new and more vitriolic turn. Opponents of the legislation have suggested the need for people to arm themselves against civil disorder—an inflammatory invitation to help break down the law and order—and have implied that enactment of the bill would stop this "essential" process from taking place. The premise is destructive, and the conclusion irrational . . .

The time for enactment of this badly needed legislation is now, before one more senseless death occurs with a cheap mail-order weapon . . .

We have a responsibility to the victims of crime and violence. It is a responsibility to think not only of our own convenience but of the tragedy of sudden death.

It is a responsibility to put away childish things, to

make the possession and use of firearms a matter undertaken only by serious people who will use them with the restraint and maturity that their dangerous nature deserves—and demands. For too long, we have dealt with these deadly weapons as if they were harmless toys. Yet their very presence, the ease of their acquisition, and the familiarity of their appearance have led to thousands of deaths each year—and to countless other crimes of violence as well . . .

It is past time that we wipe this stain of violence from our land.

The NRA and its allies killed the legislation. Less than a year later, Robert Kennedy died from three shots fired from a cheap handgun. Afterward, following floor debates in which many members of Congress evoked Kennedy's memory, aspects of the bill were resurrected and passed; over the years, the NRA succeeded in weakening them yet again. Such bills still face seemingly insurmountable opposition, and fifty years after his death, deadly weapons are far more prevalent and, on the whole, less regulated than when he testified his warnings.

Foreign Policy

Robert Kennedy had begun his public career writing and speaking about foreign affairs, befitting his heritage as an ambassador's son who enjoyed from early childhood the luxury of frequent travel abroad. Both his role as a Senate committee counsel in the 1950s and most of his duties as attorney general in his brother's administration focused on domestic affairs, but Kennedy's interest in foreign policy never slackened, and he undertook substantial and nearly continuous assignments regarding the world's trouble spots at President Kennedy's request.

Similarly, although most of his speeches while a senator concerned domestic policy, Kennedy continued to travel extensively abroad and to seek the counsel of visiting foreign dignitaries, opposition leaders, journalists, artists, students, and others. He felt a particular responsibility to his brother's international legacy, and he strove to advance the philosophy identified with Kennedy administration programs such as the Alliance for Progress: greater cooperation among equals and an emphasis on adherence to democracy and economic advancement, over expressions of Cold War allegiance or protection of U.S. business interests.

This section presents Robert Kennedy's addresses in the Senate on the dangers of nuclear weapons (June 1965) and on social conditions in Latin America (May 1966). Three of the major speeches during Kennedy's extraordinary trip to South Africa in June 1966 follow. America's traditions and role in the world are the subject of the section's final excerpt, from an appearance in October 1966.

Controlling the Spread of Nuclear Weapons

UNITED STATES SENATE

JUNE 23, 1965

It was not the danger posed by handguns in our homes, by poverty along our country back roads, or by despair in our city streets—or even by America's inexorable march to folly in the rice paddies of Vietnam—but the threat posed by the proliferation of the ultimate weapons of the twentieth century that Robert Kennedy chose as the subject of his formal maiden speech in the United States Senate on June 23, 1965. In this, Kennedy was clearly influenced by his deep involvement in the Cold War's closest brush with Armageddon: the Cuban Missile Crisis in October 1962. On June 10, 1963, in his seminal speech at American University, John Kennedy had warned of "one of the greatest hazards which man faces: . . . the future spread of nuclear arms." A month and a half before his death, President Kennedy signed the Nuclear Test Ban Treaty, calling it "a small step towards safety." His brother was determined to see the country continue on the journey.

Robert Kennedy had risen on the floor of the Senate before, including in response to the American invasion of the Dominican Republic, but he respected the institution's tradition that freshmen not present a formal address until they had acclimated to the body. Six months after being sworn in, he was ready.

Kennedy never seriously considered another choice of topic, and he solicited an initial draft from his friend Frederick G. Dutton, who had been an assistant secretary of state. His two legislative aides, Edelman and Walinsky, short on foreign policy experience but long on chutzpah, told their boss at a Friday office cocktail party (only days before he was scheduled to speak) that they found Dutton's work "a disaster." Kennedy gave instructions for a new direction and content, and Walinsky spent the weekend getting a crash course in the intricacies of nuclear policy. Kennedy suggested Walinsky speak with Roswell Gilpatric, his brother's deputy secretary of defense; and Averell Harriman, who had served in two senior State Department posts in the Kennedy administration and had been the principal United States architect of the Nuclear Test Ban Treaty.

Gilpatric later remembered Kennedy's working style when the two met to discuss the draft at Washington's National Airport:

> *First the Kennedy convertible drove up, filled with Bob, Ethel, three or four children . . . plus Brumus [the family's large Newfoundland]. We all repaired to the Admirals Club, where Bob*

and I went through the latest draft of his speech while . . . the rest of the party nearly succeeded in dismantling the Admirals Club. Bob's powers of concentration were unparalleled. Obviously he had thought out the whole complex subject of nonproliferation and knew precisely where he came out and what he wanted to get across.

As modern viewers of C-SPAN know, most Senate speeches play to empty chambers. But when Robert Kennedy rose for his first address, more than half of his colleagues had turned out to hear him.

I rise today to urge action on the most vital issue now facing this nation and the world. This issue is not in the headlines. It is not Vietnam, or the Dominican Republic, or Berlin. It is the question of nuclear proliferation—of the mounting threat posed by the spread of nuclear weapons . . .

Nuclear capability . . . will soon lie within the grasp of many. And it is all too likely that if events continue on their present course, this technical capability will be used to produce nuclear weapons . . .

Once nuclear war were to start, even between small, remote countries, it would be exceedingly difficult to stop a step-by-step progression of local war into a general conflagration.

Eighty million Americans, and hundreds of millions of other people, would die within the first twenty-four hours of a full-scale nuclear exchange. And as Chairman Khrushchev once said, the survivors would envy the dead.

This is not an acceptable future. We owe it to ourselves, to our children, to our forebears and our posterity, to prevent such holocaust. But the proliferation of nuclear weapons immensely increases the chances that the world might stumble into catastrophe.

President Kennedy saw this clearly. He said, in 1963, "I ask you to stop and think what it would mean to have nuclear weapons in so many hands, in the hands of countries large and small, stable and unstable, responsible and irresponsible, scattered throughout the world. There would be no rest for anyone then, no stability, no real security, and no chance of effective disarmament." . . .

There could be no security—when a decision to use these weapons might be made by an unstable demagogue, or by the head of one of the innumerable two-month governments that plague so many countries, or by an irresponsible military commander, or even by an individual pilot. But if nuclear weapons spread, they may be thus set off, for it is far more difficult and expensive to construct an adequate system of control and custody than to develop the weapons themselves . . .

Think just of the unparalleled opportunities for mischief: A bomb obliterates the capital city of a nation in Latin America, or Africa, or Asia—or even the Soviet Union, or the United States. How was it delivered: By plane? By missile? By car or ship? There is no evidence. From where did it come: A jealous neighbor? An internal dissident? A great power bent on stirring up trouble? Or an anonymous madman? There is only speculation. And what can be the response—what but a reprisal grounded on suspicion, leading in ever-widening circles to the utter destruction of the world we know.

The need to halt the spread of nuclear weapons must be a central priority of American policy. Of all our major interests, this now deserves and demands the greatest additional effort. This is a broad statement, for our interests are broad . . . And the crises of the moment often pose urgent questions, of grave importance for national security. But these immediate problems, and others like them, have been with us constantly for twenty years—and will be with us far into the future. Should nuclear weapons become generally available to the world, however, each such crisis of the moment might well become the last crisis for all mankind.

Thus none of the momentary crises are more than small parts of the larger question of whether our politics can grow up to our technology. The nuclear weapon, as Henry Stimson said, "constitutes merely a first step in a new control by man over forces of nature too revolutionary and dangerous to fit into the old concepts . . . It really caps the climax of the race between man's growing technical power for destructiveness and his psychological power of self-control and group control—his moral power." . . .

We cannot allow the demands of day-to-day policy to obstruct our efforts to solve the problem of nuclear spread. We cannot wait for peace in Southeast Asia, which will not come until nuclear weapons have spread beyond recall. We cannot wait for a general European settlement, which has not existed since 1914. We cannot wait until all nations learn to behave, for bad behavior armed with nuclear weapons is the danger we must try to prevent.

Rather, we must begin to move *now,* on as many fronts as possible, to meet the problem . . .

I therefore urge immediate action along the following lines.

First: We should initiate at once negotiations with the Soviet Union and other nations with nuclear capability or potential, looking toward a nonproliferation treaty. This treaty would bind the major nuclear powers not to transfer nuclear weapons or weapons capability to nations not now in possession of them. And it would pledge nations without nuclear arms, on their part, not to acquire or develop these weapons . . .

Second: We should immediately explore the creation of formal nuclear-free zones of the world . . . [where] the nuclear powers pledge not to introduce any nuclear weapons into these areas, the nations of the areas pledge not to acquire them, and appropriate machinery for the verification of these pledges is set up . . .

Third: We should complete the partial test-ban agreement of 1963 by extending it to underground as well as above-ground tests . . .

Fourth: We should act to halt and reverse the growth of the nuclear capabilities of the United States and the Soviet Union, both as to fissionable material for military weapons purposes and as to the strategic devices to deliver such material. Freezing these weapons at their present levels—which, as we all know, are more than adequate to destroy all human life on this earth—is a prerequisite to lowering those levels in the future.

The young correspondent for the *Boston Post* on King David Street in Jerusalem, March 1948, as the British mandate was ending and fighting between Arabs and Jews escalated. Behind him is a British military checkpoint at the intersection of what is today Agron Street. *John F. Kennedy Presidential Library*

With U.S. Supreme Court Justice William O. Douglas in native costume in Stalinabad (now Dushanbe, Tajikistan), Soviet Central Asia, August 1955. *RFK Human Rights*

Leaving the Senate Rackets Committee hearings with brother John, June 1957. Look/*Doug Jones*

During a break in the Senate Rackets Committee hearings in 1957, the Kennedy brothers examine one type of toy truck sold by Dave Beck Jr. that the Teamsters boss Dave Beck Sr. told union officials to "buy or you'll answer to me." Look/*Doug Jones*

Before the March 1960 New Hampshire primary, John Kennedy's presidential campaign manager is interviewed by a major Boston radio station. *Fred Forbes*

John, Robert, and Edward Kennedy in Hyannis Port, Massachusetts, July 1960. *John F. Kennedy Presidential Library*

Campaign manager surrogates for his brother in Syracuse, New York, 1960. *Truxton Hosley*

History's second-youngest U.S. attorney general in an undated portrait. *National Archives*

With Ethel, departing D.C. for their twenty-six-day goodwill tour around the world, February 1962. *USIA*

Protestors, ahead of the Kennedys' arrival in Tokyo, February 7, 1962. *USIA*

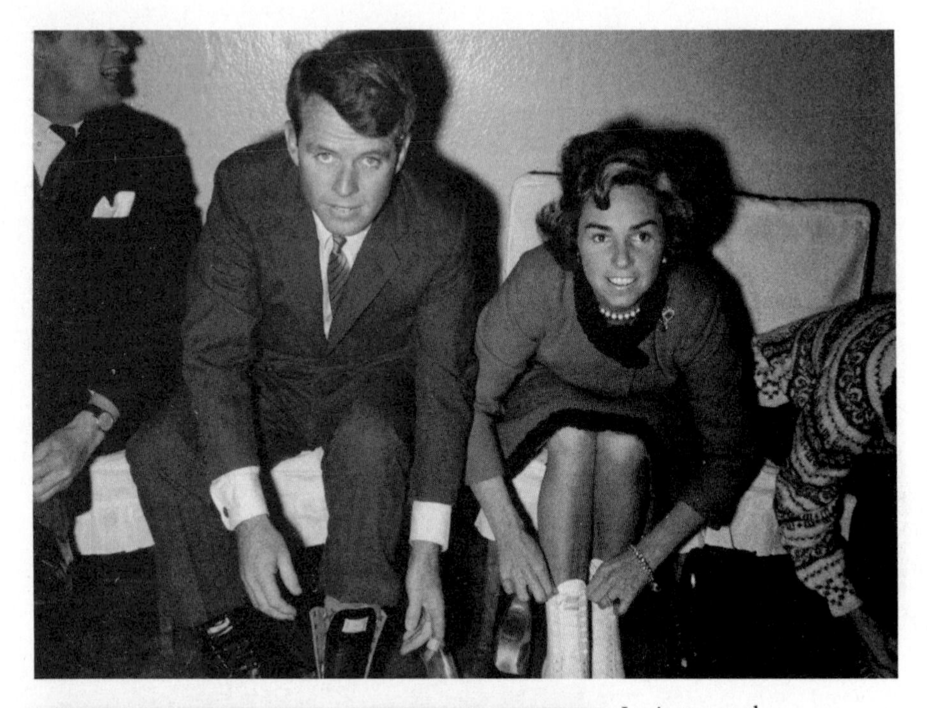

Lacing up at the Korakuen Ice Palace in Tokyo, February 1962. *USIA*

Playing in Japan, February 1962. *USIA*

Speech at the University of Indonesia, Jakarta, February 14, 1962. *USIA*

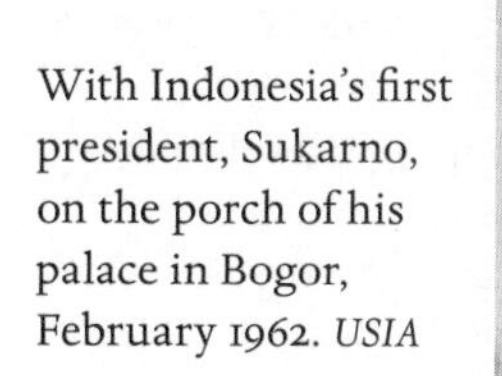

With Indonesia's first president, Sukarno, on the porch of his palace in Bogor, February 1962. *USIA*

Escorted by host Balinese nobleman Tjokorde Agung on a tour of the grounds of a Hindu temple in Ubud, Bali, February 19, 1962. *USIA*

On a freezing day in February 1963, the attorney general and his press secretary and friend, Ed Guthman, attempt their fifty-mile hike. Kennedy's inscription reads: "For Ed Guthman—This is the beginning of the trip and you are looking remarkably well—comparatively. Congratulations to an old man." *Jim Mahan/Ed Guthman*

With Ed Guthman in the attorney general's office at the Department of Justice, early 1963. (Guthman had broken his arm at an ice-skating party.) Note the Kennedy children's drawings taped to the wall over RFK's right shoulder. *Edwin Guthman*

Flanked by Dr. Martin Luther King Jr. and the NAACP's Roy Wilkins, the attorney general joins Vice President Johnson and others in the Rose Garden after President Kennedy's meeting with civil rights leaders, June 22, 1963. *Abbie Rowe, White House photographer*

Addressing the Platform Committee of the 1964 Democratic National Convention.

Readying for his Amazon adventure in Brazil, November 1965. *Steve Schapiro*

With Albert Lutuli, banned African National Congress president and 1960 Nobel Peace Prize winner, at Chief Lutuli's home in exile outside of Durban, South Africa, June 8, 1966. *USIA*

Soweto township, Johannesburg, South Africa, June 8, 1966. *USIA*

Carrying daughter Kerry on a break during the family's white water rafting trip on the Middle Fork of the Salmon River, Idaho, summer 1966. *George T. Henry*

Another private moment with Kerry on the Salmon River trip, 1966. *George T. Henry*

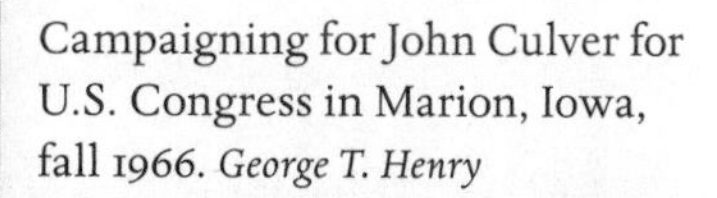

Campaigning for John Culver for U.S. Congress in Marion, Iowa, fall 1966. *George T. Henry*

Visiting the Bedford-Stuyvesant neighborhood in Brooklyn, New York. With Kennedy's help, the Restoration Corporation there combined "the best of community action with the best of the private enterprise system," improving lives for tens of thousands since 1967.
Dick DeMarsico

Robert and Ethel Kennedy at their Hickory Hill home with their (then) ten children in the late fall of 1967.
Steve Schapiro

Calling for recapturing America's moral vision, at the Allen Fieldhouse of the University of Kansas, March 18, 1968. Supporters of potential Republican presidential candidate Nelson Rockefeller (then governor of New York) hoist a sign from the balcony. *Kansas State Historical Society*

Migratory Labor Subcommittee hearings at Old Stockton High School, California, March 24, 1968. *Al Golub*

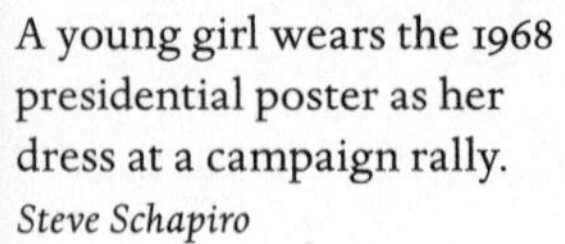

A young girl wears the 1968 presidential poster as her dress at a campaign rally. *Steve Schapiro*

With the child of migrant workers in a dry riverbed in California's Central Valley, 1968. *(c) 1976 George Ballis/TakeStock*

Breaking the fast with Cesar Chavez in Delano, California, March 10, 1968. *(c) 1976 George Ballis/ TakeStock*

Memorial Day, May 30, 1968, on the campaign train in Modesto, while barnstorming through California's Central Valley with wife, Ethel, and photographer Bill Eppridge. *Al Golub*

Revising a speech draft on the 1968 campaign plane. *Steve Schapiro*

Moreover, as [Defense] Secretary [Robert] McNamara has shown, it would be in the direct self-interest of the United States and the Soviet Union to cut back our nuclear forces. For we each have more than enough to destroy the other nation, yet can never acquire enough to prevent our own destruction . . .

Fifth: We should move to strengthen and support the International Atomic Energy Agency . . .

Sixth: It is vital that we continue present efforts to lessen our own reliance on nuclear weapons. Since 1961, we have worked to build up our nonnuclear forces, and those of our allies, so that if conflict comes, we need not choose between defeat and mutual annihilation . . .

As to all these points—in all our efforts—we will have to deal with one of the most perplexing and difficult questions affecting American foreign policy: China. It is difficult to negotiate on any question with the intransigent leaders of Communist China. And it is doubly difficult when we are engaged in South Vietnam. China is profoundly suspicious of and hostile to us—as we are highly suspicious of her. But China is there. China will have nuclear weapons. And without her participation it will be infinitely more difficult, perhaps impossible in the long run, to prevent nuclear proliferation . . .

At an appropriate time and manner, therefore, we should vigorously pursue negotiations on this subject with China. But if we must ultimately have the cooperation of China, and the Soviet Union, and France, and all other nations with any nuclear capability whatever, it does not follow that we should wait for that cooperation before beginning our efforts. We are stronger, and therefore have more responsibility, than any nation on earth; we should make the first effort—the greatest effort, and the last effort—to control nuclear weapons. We can and must begin immediately . . .

And we can and must continue to reexamine our own attitudes, to ensure that we do not lapse back into the fatalistic and defeatist belief that war is inevitable, or that our course is too fixed to be affected by what we do—to remember, as President Kennedy said, that "no government or social system is so evil that its people must be considered as lacking in virtue," and to remember that "in the final analysis, our most basic common link is that we all inhabit this small planet. We all breathe the same air. We all cherish our children's future. And we are all mortal."

Above all, we must recognize what is at stake. We must face realities—however unpleasant the sight, however difficult the challenge they pose us. And we must realize that peace is not inaction, nor the mere absence of war. "Peace," said President Kennedy, "is a process—a way of solving problems." It is only as we devote our every effort to the solution of these problems that we are at peace; it is only if we succeed that there will be peace for our children.

The reaction to Kennedy's speech was considerable and immediate, and, from the White House, not favorable. Johnson saw, especially in the litany of quotes from his predecessor, an implicit criticism of his administration's inability to advance on superpower arms reduction, nonproliferation, or elimination of testing. Columnist Drew Pearson, a frequent recipient of White House leaks, charged that Kennedy's timing was deliberately aimed at upstaging Johnson's address to the United Nations two days later, and that Kennedy had ignored two emissaries sent by Johnson with pleas to forbear in the national interest. The story was flatly refuted by Kennedy's people. (Walinsky has asserted that he reviewed the speech in detail with the President's national security adviser prior to delivery.) Whatever the background, Johnson removed disarmament proposals from his United Nations address, to avoid any impression of following his rival's lead.

Seventeen of Kennedy's colleagues praised his remarks from the floor. Kennedy looked to the press for an early reaction: Walinsky remembered a wake-up call the morning after, from an irate Kennedy, angered that the *Washington Post* had buried its coverage in the middle of the paper and characterized the speech as a rehash of existing administration proposals. Walinsky responded by asking his boss how the *New York Times* carried the story. Kennedy hadn't received his copy of the paper, so Walinsky left the phone to retrieve his.

> *There was the greatest relief of my life up to that point, because the thing was spread all over the front page and they had a picture of him . . . and they reprinted almost the whole text . . . something that newspapers . . . just don't do for senators' speeches . . . So then [Kennedy] said, "All right. Well, that's okay. So that's just the* Washington Post. *They're just screwing around with Johnson. The hell with them."*

Public reaction was substantial, with more than ten thousand letters pouring into the office, nearly all favorable and many thanking Kennedy for awakening them to an underappreciated danger.

The subject remained a critical one for Kennedy. He delivered a second policy address on the subject in the

Senate, and his last press secretary, Frank Mankiewicz, remembered that "there was hardly an extemporaneous speech that [Kennedy] ever gave [during the 1968 campaign] that he didn't return to that theme: that if we didn't bring nuclear weapons under control, our children are all going to die."

Redirecting United States Policy in Latin America

UNITED STATES SENATE

MAY 9 AND 10, 1966

Stopping the proliferation of nuclear weapons was but one of the interests that carried over from Kennedy's executive branch tenure to his Senate agenda. He had developed a fascination with Latin America while serving in his brother's administration. In fact, Dick Goodwin, who wrote speeches for both Kennedys and for Lyndon Johnson, believed that "had President Kennedy lived, his intention was to move Bobby from the Justice Department and put him in charge of Latin American affairs."

The region received more attention from the Kennedy administration than from any of its predecessors; particularly appreciated was the United States' economic aid, which was ten times greater in John Kennedy's three years than it had been in the previous sixteen years. The United States had had an overpowering presence in the Americas for more than a century, traditionally supporting (and occasionally installing) regimes that would best serve the interests of United States mineral and commodity extractors—and, as the Cold War became the prism through which the nation's entire foreign policy was viewed, the interests of Communist containment. Under the Kennedy administration, most of the money was aimed at providing the conceptual fulcrum for democratic economic and political development under the Alliance for Progress.[10]

It was a commitment Robert Kennedy felt his brother's successor had abandoned. In February 1965, the Alliance for Progress's first chief, Teodoro Moscoso, said plaintively, "President Kennedy stood for change—revolutionary change—and he said so even when it hurt in exalted places . . . Do we remember [now] that there is a revolution going on?" Certainly Johnson's first two steps in the region smacked of the old principles: He appointed Thomas Mann (regarded as an apologist for the old order) as the top State Department official on inter-American affairs (the head of Standard Oil of New Jersey said the appointment confirmed that the business community was once again "'in' with the United States government"), and he recognized the military junta that overthrew the elected government of Juan Bosch in the Dominican Republic. Later that spring, Johnson sent twenty-two thousand United States troops to the Dominican Republic to prop up the junta in the midst of an uprising. Reaction throughout the continent was nearly uniformly hostile, and there was widespread discussion of the collapse of the Alliance. Robert Kennedy opposed the intervention and the policy direction it represented.

Having received a speaking invitation in Brazil, Kennedy prepared for a trip to South America over the 1965 Thanksgiving congressional recess. He gave Walinsky a long list of individuals around the world with whom to consult in advance of the trip and a further instruction to him "and everybody else who was involved to keep reaching out, always reaching out, and finding more and more people."

Walinsky also noted that Kennedy immersed himself in "many, many long hours [of] briefing sessions," and that he maintained broad contacts throughout the region, "not just Peace Corps people, but priests and editors and government leaders." Far less useful was the briefing he received from the State Department: Apprised of Johnson's hostility to Kennedy's venture into foreign affairs, the meeting seemed deliberately designed to be adversarial and nonproductive.

Kennedy worried about two potentially adverse consequences of his trip. First, he expected that the bulk of the United States press would characterize any criticism of administration policy he might deliver while in Latin America as undermining the government's efforts and the national interest. Second, he worried that such criticism might be destabilizing within the countries themselves: As "heir" of the martyred president (revered in the region), Kennedy's comments carried great weight in Latin America; as the only substantial and potential Democratic rival of the sitting president, Kennedy had virtually no ability to have our government back up whatever democratic forces his words might unleash.

On November 10, 1965, Kennedy and his wife, accompanied by friends and a press contingent, left Miami. Their first stop, in Lima, Peru, set the tone for the trip: Kennedy avoided the staged events arranged by the United States embassy and spent his time with students, intellectuals, and journalists, and touring some of the poorest neighborhoods on the planet. Whenever he appeared in public, he was inundated by crowds seeking not only to listen but, in an almost spiritual manner, to touch him.

Campus appearances remained his forte. He was frank in his views on current United States policy as well as on the inevitability of the revolutionary forces for economic redistribution and political empowerment. But he also chastised those who reflexively blamed the United States for the region's ills. "You are the Peruvian leaders of the future," he told the students at Lima University. "You have to decide what is in your interests. If you object to American aid, have the courage to say so. But you are not going to solve your problems by blaming the United States and avoiding your own personal responsibility to do something about them."

After Peru, the party decamped for Chile. After meeting with government leaders, Kennedy traveled to Concepcion, Chile's third-largest city and home to a university with a large body of Marxist students, who threatened to disrupt his scheduled campus speech. Kennedy debated with them privately for two hours, and he offered to continue to do so publicly at the evening's event, but the students refused. The university's rector, fearing violence (particularly against Kennedy), asked him to withdraw, but moderate student leaders told him that to do so would be judged a Communist triumph and could ensure that no other democratic spokesman would be able to appear on campus. Kennedy decided to deliver the speech as scheduled.

Arthur Schlesinger Jr. described the scene that followed, as Kennedy entered

> *a gymnasium packed with students screaming "Kennedy—paredon" (to the wall), and throwing eggs and garbage. "It was really a frightening goddamned thing," said Martin Arnold of the* New York Times. *Eggs splattered over [his*

> *companions] . . . but Kennedy walked along, never looking back, and was untouched. "If these kids are going to be young revolutionaries," he muttered . . . "they're going to have to improve their aim."*
>
> *He tried to make himself heard from the platform. After a moment, he challenged the Marxists to debate: "Will you test your ideas before the students of this university?" . . . Some had leaned across the rails to shake his hand . . . [but one] student spat in his face. The [Marxist leader Kennedy had met with that afternoon] grabbed the spitter by his shirt and pulled him away. The pandemonium continued. Finally Kennedy left to a thunderous ovation. [He and his party] went to bed well after midnight. At four a.m. [friend and traveling companion William] vanden Heuvel roused the party: "Come on, the senator wants to go to the coal mines."*

Coal miners in the area, predominantly Communists, earned $1.50 a day for backbreaking, dangerous work in tunnels stretching out for miles under the ocean floor. Chilean security forces were appalled by the risk, and as Kennedy and his friends started out, they saw soldiers stationed every hundred yards along the road to the mines. Kennedy descended by a small open elevator fifteen hundred feet into the earth, then took a series of coal trains underground for five miles, through low-hanging high-voltage wires. "All these coal miners were wild to see him," the *Times*'s Arnold remembered. "If I worked in this mine," Kennedy told a companion, "I'd be a Communist too."

Next, Brazil, where Kennedy celebrated his fortieth birthday. Two days later, on the second anniversary of his brother's assassination, he toured the seaside slum of Recife, where the stench was so foul that his Brazilian security escorts would not leave their cars. One hundred thousand people came out to cheer him in a four-hour motorcade that followed. With a few companions, Kennedy chartered an amphibious plane and flew deep into the Amazonian jungle. There, Kennedy spent a day worthy of Indiana Jones. First, an incident related by Andrew Glass of *The Saturday Evening Post,* traveling with Kennedy to an inland lake:

> *A wide, low wagon was hitched to a tractor, and we hopped aboard. Kennedy invited dozens of children running beside . . . the already overloaded flatbed to climb on . . . The tractor at one point swerved suddenly to avoid a deep hole . . . Kennedy was hurled off the flatbed and*

> *landed in the bushes on his back. He sprang to his feet, borrowed a bicycle from a boy riding behind the tractor, and pedaled the two remaining miles to the lake while carrying two children on the bike's rear fender. At the lake, still surrounded by curious children, Kennedy swam out to a large log bobbing thirty yards offshore and tossed a football . . . with Goodwin. Not until we left did the local officials tell us that the lake was really a branch of the Solimoes and consequently infested with flesh-eating piranha fish.*

A dusk fishing excursion by canoe in a tropical deluge followed (the result: four fish, despite the shore-bound native guides' prediction that the three gringos would never return, venturing with no visibility into a trackless jungle). To stay on schedule for Venezuela, Kennedy arose at three thirty the next morning and, with four guides, paddled a dugout canoe downriver fifteen miles to a spot that could accommodate the plane. Companion vanden Heuvel remembered that "when the canoe foundered in the rapids, Kennedy jumped overboard and pushed the boat forward. Splashing along in the river, he mimicked [legendary CBS anchorman] Walter Cronkite in declaring: 'It was impossible to pinpoint the exact time and place where he decided to run for president. But the idea seemed to take hold as he was swimming in the Amazonian river of Nhamunda, keeping a sharp eye peeled for man-eating piranhas. Piranhas have never been known to bite a U.S. senator.'"

Kennedy waited five months to present his conclusions about Latin America. Other pressing matters, particularly the escalation of the American effort in Vietnam, commanded his attention on his return, and he wanted the time to consult as widely regarding his policy suggestions as he had on his trip planning. The result was the longest speech of his career, spanning two Senate sessions (May 9 and 10, 1966). Kennedy began torturously, trying to avoid a public break with the president by suggesting that his views were a reflection of Johnson's goals rather than critical of them. He continued:

But before discussing specific problems, there is one element of our policy that must be clear: . . . That we associate ourselves with the aspirations of the Latin American people for a better life, for justice between men and nations, for the dignity of freedom and self-sufficiency. These demands are in part material; above all, they are demands of the spirit.

But we must realize that the demands of the spirit—the demands for justice and a sense of participation in the life of one's country—are the essential preconditions to material progress . . .

Five years ago . . . President Kennedy called on all the people of the hemisphere to join in a new Alliance for Progress . . . It was far more than a promise of economic development. In addition, it pledged: a more equitable distribution of national incomes . . . diversification of national economic structures . . . of industrialization . . . comprehensive agrarian reform . . . the elimination of illiteracy . . . improved health . . . expanded housing and public services . . .

The coal miners in Concepcion, laboring five miles under the sea for $1.50 a day; the mothers in Andean villages where schoolteachers tell the children that their parents' tongue is the speech of animals; the cane cutters and laborers watching their children die; the priests who see the teachings of their church violated by the lords of the land; these people are the engines of change in Latin America.

These people will not accept this kind of existence for the next generation. We would not; they will not. There will be changes.

So a revolution is coming—a revolution which will be peaceful if we are wise enough; compassionate if we care enough; successful if we are fortunate enough—but a revolution which is coming whether we will it or not. We can affect its character; we cannot alter its inevitability.

But to say this is only the beginning; the question is now how the revolution is to be made and guided.

At the heart of the revolution, underlying all hope for economic progress and social justice, are two great and resistant problems: education and land reform. Both education and land reform are needed for economic growth. No amount of capital, no purely economic measures, can bring progress unless each nation has the trained and skilled people to do the work of modernization and change. Nor can any industrial economy be built on a failing, inadequate, and obsolete system of agricultural production.

But these are far more than economic measures. No matter how rich or powerful a nation may grow, children condemned to ignorance, families enslaved to land they cannot hope to own, are denied the dignity—the fulfillment of talent and hope—which is the purpose of economic progress. Progress without justice is false progress and a false hope . . .

I know that, ever since the onset of the Cold War, we have been urged to develop a concise, exciting American manifesto—a platform which would compete with the simple rousing calls of the Communists. But what matters about this country cannot be put into slogans; it is a process, a way of doing things and dealing with people, a way of life. There are two major ways of telling others what this country is really about: to bring people here or to send Americans abroad . . .

I believe that we are ready to recognize that foreign aid is not a "giveaway"—rather, that it is both a moral obligation to fellow human beings and a sound and necessary investment in the future . . .

And I believe we are ready to recognize that millions saved now can mean billions lost five or ten or twenty years from now, and that the human cost of delay is incalculable. Time after time, in these uncertain and dangerous years, we have reaped the consequences of neglect and delay, of misery and disease and hunger left too long to fester unremedied . . . As President Kennedy said, "If we cannot help the many who are poor, we cannot save the few who are rich." . . .

There is temptation—one to which we have sometimes given in—to use our great power and our aid to force agreement from other nations, or to punish them for their disagreement. This temptation is most obvious in matters of foreign policy and the Cold War; whether a country voted in the OAS to approve our action in the Dominican Republic, or perhaps whether it recognizes Communist China or votes for its admission to the United Nations. These are matters of considerable importance in the United States. It is understandable that officials in the executive branch or members of Congress or others in the country would feel that nations which fail to stand with us are not reliable allies and should not receive U.S. assistance.

But this feeling, so understandable in the passions and excitement of the moment, can only be harmful over the long run. We expect our government to reflect the feelings of our people. Latin Americans expect the same from their governments and deeply resent any government which seems less than fully independent in its decisions. One Latin American president put it to me succinctly: "If you want a government that says always, 'yes, yes, yes,'" he told me, "you will soon have to deal with a government that says always 'no, no, no.'" The forces of progressive democracy in Latin America have many enemies and obstacles. They are under attack from the right and the left—from those who would sacrifice justice to preserve the past and from those who would heedlessly impose bloodshed and dictatorship to hasten change.

. . . That is the great danger of subversion in Latin America: that if we allow Communism to carry the banner of reform, then the ignored and the dispossessed, the insulted and injured, will turn to it as the only way out of their misery.

And I regret to say that some of our actions, and those of some Latin American governments, have helped the Communists to do this. For many years, the established order in Latin America has referred to all efforts for reform and justice as Communist . . .

This we must clearly understand. Communism is not a native growth in Latin America. Given any meaningful alternative, its people will reject Communism and follow the path of democratic reform and membership in the inter-American system. But if we allow ourselves to become allied with those to whom the cry of "Communism" is only an excuse for the perpetuation of privilege—if we assist, with military materials and other aid, governments which use that aid to prevent reform for their people—then we will give the Communists a strength which they cannot attain by anything they themselves might do.

President Johnson's Mexico City address set what must be our policy: "We will not be deterred," he said, "by those who say that to risk change is to risk Communism." . . .

The greatest success for nations, as for individuals, is found in truth to themselves. We did not build the United States on anti-Communism. The strength of our institutions, the energy and talent of Americans, come out of our long struggle to build a nation of justice and freedom and happiness. Ours is the strength of positive faith; we need neither to hate nor fear our adversaries. Let our emphasis be, then, less on what the Communists are doing to threaten peace and order in Latin America, and more on what we can do to help to build a better life for its people . . .

Leadership in freedom cannot rest on wealth and power. It depends on fidelity and persistence in those shaping beliefs—democracy, freedom, justice—which men follow from the compulsions of their hearts and not the enslavement of their bodies. We must cope with real dangers, overcome real obstacles, meet real needs; but always in a way which preserves our own allegiance to the principles of the Alliance. Otherwise we will preserve the shadow of progress and security at the expense of the substance of freedom in the New World.

Kennedy's trip and his Senate speech received substantial attention in Latin America and in the United States press but had no impact on administration policy. The atrophying of the Alliance, and the return to perpetuating the status quo, accelerated as Vietnam consumed Johnson's foreign policy.

Kennedy continued his interest in Latin America, meeting regularly with visitors, in and out of government, from the region. The squandered opportunities of the Alliance, the depth of the hostility of the Chilean Communist students toward the United States, and the extreme poverty throughout the continent all found frequent expression in his speeches, particularly his extemporaneous ones.

Robert Kennedy in South Africa

Day of Affirmation

UNIVERSITY OF CAPE TOWN

JUNE 6, 1966

As he summarized for his Senate colleagues the conclusions drawn from his travels in South America, Kennedy readied himself for a longer journey to another continent. In the fall of 1965, Ian Robertson, a twenty-one-year-old medical student who was president of the National Union of South African Students (NUSAS), invited Kennedy to be the keynote speaker at the annual Day of Reaffirmation of Academic and Human Freedom, to be observed at the University of Cape Town in the spring of 1966.

Kennedy was no stranger to African affairs. He had headed an American delegation to the Ivory Coast in August 1961, when that nation celebrated its first anniversary of independence; and as a member of the National Security Council, he participated in critical decisions concerning the continent during his brother's presidency. Perhaps more important, he functioned as a back channel for dissident viewpoints.

G. Mennen Williams, under secretary of state for African affairs under Presidents Kennedy and Johnson, recalled that during the Kennedy administration, Robert Kennedy

> *had both a significant interest and a significant impact [on United States policy toward Africa] . . . The president shifted a lot of people to Bobby that he couldn't see himself . . . There were some people that had a more revolutionary task that [the president] couldn't very well diplomatically see. But Bobby saw them and helped [his brother], helped [the dissident Africans] through overt and covert means and was sympathetic and helpful.*

It was a role guaranteed to displease the government of South Africa.

As attorney general, Kennedy did not inform the South Africans when he met in Washington with Patrick Duncan, an opponent of South Africa's policy of apartheid. Foreign Minister Eric Louw filed a formal protest with the U.S. government, and his government placed Duncan in internal exile (called *banning),* making it illegal for him to be in a room with more than one person at a time or to be quoted in the press in any way. Such harsh treatment was consistent with the South African government's methods for preserving white control in a nation that, in 1966, had three million whites, twelve million blacks, and two million persons of Indian descent or multiracial heritage (whom the South African government classified as *coloureds*). During the remainder of his tenure at the Justice Department, Kennedy continually checked with the State Department to ensure that contact was maintained with Duncan and all possible assistance provided to him.

In an article he wrote for *Look* magazine summarizing his 1966 trip, Kennedy described for his American audience the conditions faced by blacks in South Africa:

> *You cannot participate in the political process, and you cannot vote. You are restricted to jobs for which no whites are available. Your wages are from 10 to 40 percent of those paid a white man for equivalent work. You are forbidden to own land except in one small area. You live with your family only if the government approves . . . You are, by law, inferior from birth to death. You are totally segregated, even at most church services.*

Kennedy wanted to accept the NUSAS invitation, to learn about the country and its people firsthand and to encourage the forces for equal justice there, particularly among the young. The South African government delayed granting a visa for five months, finally relenting in March 1966—and then, only for a trip limited to four days, and in June rather than May, as requested. The official attitude toward the visit remained unremittingly hostile. First, Kennedy was informed he would not be invited to meet with any governmental ministers. Next, three weeks before Kennedy's arrival, the government "banned" his host, Ian Robertson, for five years, for inviting Kennedy. And finally, six days before Kennedy's scheduled arrival, Pretoria denied travel permits for forty foreign journalists, mainly from the United States, assigned to the trip.

To prepare for his journey, Kennedy read widely, followed the congressional hearings held in the spring on United States–South African relations, and scheduled regular briefing sessions, some at Hickory Hill, with academics and others. Of particular importance was the prodding of Allard Lowenstein.[11] Having risen to public prominence after addressing the United Nations following his clandestine 1959 trip to South West Africa (now Namibia) to collect testimony about conditions under apartheid in South Africa, Lowenstein reviewed an early draft of the Cape Town address that Walinsky later described as "much more delicately phrased [and] nonrevolutionary," aiming to avoid antagonizing whites. Lowenstein, who had become a prominent

U.S. civil rights activist, convinced Kennedy and his staff that it was essential to confront South Africa's institutionalized racism directly and dramatically. Dick Goodwin was enlisted to help with the redrafting.

Stopping in London en route, Kennedy met with a small group of people experienced in African affairs. Anthony Lewis, then the London bureau chief for the *New York Times,* remembered that most members of the group "were deeply engaged in opposition to Afrikaner racism, and they tended to suggest sweeping emotional positions." Kennedy pressed for something deeper, telling them, "It's easy to make speeches that would win votes in New York, but that won't help the Africans—it might even be harmful. What can I say that will really be useful to them?"

South Africa's papers were full of speculation about Kennedy's intentions: the pro-government press branded him a troublemaker who came to consort with what the government called "a damnable and detestable organization . . . a breeding nest of vipers," and even the English-language *Rand Daily Mail* noted that "there are those who say that his purpose is purely to advance his political ambitions at home, that he wishes to capture the Negro vote in his campaign for the presidency by championing the cause of the black man in the last outposts of white domination."

Despite official hostility, widespread uncertainty about his motives, and a government curfew that subjected black attendees to arrest, a largely white crowd of more than fifteen hundred waited until nearly midnight, on Saturday, June 4, 1966, to greet Robert and Ethel Kennedy at Johannesburg's Jan Smuts Airport. It was a unique venue, but a typical Kennedy welcome: The screaming, surging crowd tore off his cufflinks and cheered his brief remarks.

Kennedy attempted to set the tone for the visit there by telling the airport gathering:

> *We are, like you, a people of diverse origins. And we too have a problem, though less difficult than yours, of learning to live together, regardless of origins, in mutual respect for the rights and well-being of all our people . . . I come here to South Africa to exchange views with you—with all segments of South African thought and opinion—on what we together can do to meet the challenges of our time . . . I hope, above all, to learn; and I thank you for giving me the opportunity to do so.*

Sunday morning, at the American embassy in the nearby capital of Pretoria, Kennedy met with opposition political leaders. Helen Suzman, a member of Parliament and leading critic of apartheid, commented afterward to a *Rand Daily Mail* reporter that Kennedy "had obviously gone to the trouble of getting very 'genned-up' (prepared) on South African affairs and his questions were pertinent . . . My main impression was that he was very interested in getting information about our country and had not come here to put down a line on how we should solve our problems."

Prevented by Prime Minister Hendrick Vetwoerd from meeting any cabinet ministers, Kennedy scheduled a lengthy session with Afrikaans newspapers' editors, to debate apartheid with its proponents. The editor of a leading pro-government daily, *Die Transvaler,* declined, saying that Sunday was not a proper day for a political discussion. The paper's backup representative presented Kennedy with a book titled *The Principle of Apartheid;* without hesitation, Kennedy then produced his own gifts: copies of his book *The Pursuit of Justice* and a hefty volume captioned *American Negro Reference Book.*

The remainder of Sunday's activities included meetings with editors of English-language papers, attendance at Mass, and numerous forays into public. According to the *Cape Times:*

After a "brunch" at the home of the American ambassador . . . [the Kennedys] suddenly appeared from a side entrance and set off on an unscheduled walk around the plush Pretoria suburb of Waterkloof. A fitness fan and originator of the "50 mile hike," Senator Kennedy set a cracking pace. All along his route he stopped to talk with Africans—most of them local servants taking a break on the grassed sidewalks—greeting them with "I'm Robert Kennedy from the United States and this is my wife, Ethel." Many of them were bewildered by the sudden attention of a large party of whites and one young African he approached with an outstretched hand took fright at the battery of cameras, let out a yelp, and ran.

At that night's dinner with business leaders, Kennedy's hosts expressed puzzlement that South Africa's staunch anti-Communism did not excuse its racial policies in American eyes. "But what does it mean to be against Communism," Kennedy asked, "if one's own system denies the value of the individual and gives all power to the government—just as the Communists do?"

"You don't understand . . . our unique problems," Kennedy was told. "We are beleaguered." Kennedy had sympathy with the white Afrikaner minority's fear of change, knowing they had themselves been discriminated against by Anglo colonists in the country's early days, and they had ultimately triumphed over their fellow settlers, the indigenous tribes, and an often pitiless landscape to create Africa's economic powerhouse. But he was certain that denial of human dignity and freedom made the whites' gains temporary and the country enormously unstable, to the risk of the entire continent.

On Monday, after morning meetings in Pretoria with African journalists and clergy leaders, he flew to Cape Town, where a cheering crowd of three thousand gave him what a local columnist described as "a real hero's welcome." Twice after his late-afternoon arrival, Kennedy made unscheduled trips into the black townships, while his wife, Ethel, toured a student-supported facility for blacks and made impromptu visits to homes of the poor in the area.

Also unscheduled was a twenty-minute visit with Ian Robertson; the two began the meeting by jumping up and down on the apartment floors to disable the authorities' bugging devices (a trick Kennedy remembered from his rackets investigations days). Kennedy presented the student leader with a copy of John Kennedy's *Profiles in Courage*, which Robert Kennedy and the president's widow, Jacqueline, had autographed. A crowd of middle-class Afrikaners had gathered outside, and they warmly cheered Kennedy's exiting wish for a swift end to Robertson's "troubles with the South African government."

He had chosen the University of Cape Town, source of his initial invitation to South Africa, as the site of his first major address, an attempt to weave appreciation for the history and legitimate aspirations of the country's white settlers with an appeal to rid themselves of their fears, and embrace broader and more enduring human traditions and mutual goals. Tension had been high for some days, and a large banner protesting Robertson's banning had been burned by unknown parties two days previous. Interest in Kennedy's speech was tremendous: Tickets for seats in the sixteen-hundred-seat hall had long been sold out, and overflow arrangements were made for more listeners, via loudspeakers, on the lawn outside and in neighboring buildings.

Outside, eighteen thousand gathered in bitter cold and wind, many for nearly three hours, before Kennedy's unmarked car made its arrival. The *Cape Times* reported that Kennedy was "immediately engulfed" by the crowd, which "was so dense he had to wait half an hour before a way was cleared." The hall itself bore banners and signs opposing the United States war effort in Vietnam. Kennedy followed a symbolic procession of school officials and students inside, led by Prunella MacRobert, identified in press reports as "the head woman student," carrying the extinguished torch of academic freedom. The audience was garbed in formal academic gowns. As he ascended to the stage, where a chair remained empty on the dais in honor of the banned Robertson, tears glistened in Kennedy's eyes.

I came here because of my deep interest and affection for a land settled by the Dutch in the mid-seventeenth century, then taken over by the British, and at last independent; a land in which the native inhabitants were at first subdued, but relations with whom remain a problem to this day; a land which defined itself on a hostile frontier; a land which has tamed rich natural resources through the energetic application of modern technology; a land which once imported slaves and now must struggle to wipe out the last traces of that former bondage. I refer, of course, to the United States of America. But I am glad to come here to South Africa. I am already enjoying my visit. I am making an effort to meet and exchange views with people from all walks of life, and all segments of South African opinion, including those who represent the views of the government. Today I am glad to meet with the National Union of

South African Students. For a decade, NUSAS has stood and worked for the principles of the Universal Declaration of Human Rights—principles which embody the collective hopes of men of goodwill all around the world . . .

This is a Day of Affirmation, a celebration of liberty. We stand here in the name of freedom.

At the heart of that Western freedom and democracy is the belief that the individual man, the child of God, is the touchstone of value, and all society, groups, the state, exist for his benefit. Therefore the enlargement of liberty for individual human beings must be the supreme goal and the abiding practice of any Western society.

The first element of this individual liberty is the freedom of speech: the right to express and communicate ideas, to set oneself apart from the dumb beasts of field and forest; to recall governments to their duties and obligations; above all, the right to affirm one's membership and allegiance to the body politic—to society—to the men with whom we share our land, our heritage, and our children's future.

Hand in hand with freedom of speech goes the power to be heard, to share in the decisions of government which shape men's lives. Everything that makes man's life worthwhile—family, work, education, a place to rear one's children and a place to rest one's head—all this depends on decisions of government; all can be swept away by a government which does not heed the demands of its people. Therefore, the essential humanity of men can be protected and preserved only where government must answer—not just to the wealthy, not just to those of a particular religion, or a particular race, but to all its people.

And even government by the consent of the governed, as in our own Constitution, must be limited in its power to act against its people; so that there may be no interference with the right to worship, or with the security of the home; no arbitrary imposition of pains or penalties by officials high or low; no restrictions on the freedom of men to seek education or work or opportunity of any kind, so that each man may become all he is capable of becoming.

These are the sacred rights of Western society. These were the essential differences between us and Nazi Germany, as they were between Athens and Persia.

They are the essence of our differences with Communism today. I am unalterably opposed to Communism because it exalts the state over the individual and the family, and because of the lack of freedom of speech, of protest, of religion, and of the press, which is the characteristic of totalitarian states. The way of opposition to Communism is not to imitate its dictatorship but to enlarge individual freedom, in our own countries and all over the globe. There are those in every land who would label as Communist every threat to their privilege. But as I have seen on my travels in all sections of the world, reform is not Communism. And the denial of freedom, in whatever name, only strengthens the very Communism it claims to oppose.

Many nations have set forth their own definitions and declarations of these principles. And there have often been wide and tragic gaps between promise and performance, ideal and reality. Yet the great ideals have constantly recalled us to our duties. And—with painful slowness—we have extended and enlarged the meaning and the practice of freedom for all our people.

For two centuries, my own country has struggled to overcome the self-imposed handicap of prejudice and discrimination based on nationality, social class, or race discrimination profoundly repugnant to the theory and command of our Constitution. Even as my father grew up in Boston, signs told him that NO IRISH NEED APPLY. Two generations later President Kennedy became the first Catholic to head the nation; but how many men of ability had, before 1961, been denied the opportunity to contribute to the nation's progress because they were Catholic, or of Irish extraction? How many sons of Italian or Jewish or Polish parents slumbered in slums—untaught, unlearned, their potential lost forever to the nation and human race? Even today, what price will we pay before we have assured full opportunity to millions of Negro Americans?

In the last five years we have done more to assure equality to our Negro citizens, and to help the deprived both white and black, than in the hundred years before. But much more remains to be done.

For there are millions of Negroes untrained for the simplest of jobs, and thousands every day denied their full equal rights under the law; and the violence of the disinherited, the insulted and injured, looms over the streets of Harlem and Watts and South Side Chicago.

But a Negro American trains as an astronaut, one of mankind's first explorers into outer space; another is the chief barrister of the United States government, and dozens sit on the benches of court; and another, Dr. Martin Luther King, is the second man of African descent to win the Nobel Peace Prize for his nonviolent efforts for social justice between races.

We have passed laws prohibiting discrimination in education, in employment, in housing, but these laws alone cannot overcome the heritage of centuries—of broken families and stunted children, and poverty and degradation and pain.

So the road toward equality of freedom is not easy, and great cost and danger march alongside us. We are committed to peaceful and nonviolent change, and that is important for all to understand—though all change is unsettling. Still, even in the turbulence of protest and struggle is greater hope for the future, as men learn to claim and achieve for themselves the rights formerly petitioned from others.

And most important of all, all the panoply of government power has been committed to the goal of equality before the law, as we are now committing ourselves to the achievement of equal opportunity in fact.

We must recognize the full human equality of all of our people—before God, before the law, and in the councils of government. We must do this, not because it is economically advantageous, although it is; not because the laws of God command it, although they do; not because people in other lands wish it so. We must do it for the single and fundamental reason that it is the right thing to do.

We recognize that there are problems and obstacles before the fulfillment of these ideals in the United States, as we recognize that other nations, in Latin America and Asia and Africa, have their own political, economic, and social problems, their unique barriers to the elimination of injustices.

In some, there is concern that change will submerge the rights of a minority, particularly where the minority is of a different race from the majority. We in the United States believe in the protection of minorities; we recognize the contributions they can make and the leadership they can provide; and we do not believe that any people—whether minority, majority, or individual human beings—are "expendable" in the cause of theory or policy. We recognize also that justice between men and nations is imperfect, and that humanity sometimes progresses slowly.

All do not develop in the same manner, or at the same pace. Nations, like men, often march to the beat of different drummers, and the precise solutions of the United States can neither be dictated nor transplanted to others. What is important is that all nations must march toward increasing freedom; toward justice for all; toward a society strong and flexible enough to meet the demands of all its own people, and a world of immense and dizzying change.

In a few hours, the plane that brought me to this country crossed over oceans and countries which have been a crucible of human history. In minutes we traced the migration of men over thousands of years; seconds, the briefest glimpse, and we passed battlefields on which millions of men once struggled and died.

We could see no national boundaries, no vast gulfs or high walls dividing people from people; only nature and the works of man—homes and factories and farms—everywhere reflecting man's common effort to enrich his life. Everywhere new technology and communications bring men and nations closer together, the concerns of one inevitably becoming the concerns of all. And our new closeness is stripping away the false masks, the illusion of difference which is at the root of injustice and hate and war. Only earthbound man still clings to the dark and poisoning superstition that his world is bounded by the nearest hill, his universe ended at river shore, his common humanity enclosed in the tight circle of those who share his town and views and the color of his skin.

It is your job, the task of the young people of this world, to strip the last remnants of that ancient, cruel belief from the civilization of man. Each nation has different obstacles and different goals, shaped by the vagaries of history and of experience. Yet as I talk to young people around the world I am impressed not by the diversity but by the closeness of their goals, their desires and their concerns and their hope for the future. There is discrimination in New York, the racial inequality of apartheid in South Africa, and serfdom in the mountains of Peru. People starve in the streets of India, a former prime minister is summarily executed in the Congo, intellectuals go to jail in Russia, and thousands are slaughtered in Indonesia; wealth is lavished on armaments everywhere in the world. These are differing evils; but they are the common works of man. They reflect the imperfections of human justice, the inadequacy of human compassion, the defectiveness of our sensibility toward the sufferings of our fellows; they mark the limit of our ability to use knowledge for the well-being of our fellow human beings throughout the world. And therefore they call upon common qualities of conscience and indignation, a shared determination to wipe away the unnecessary sufferings of our fellow human beings at home and around the world.

It is these qualities which make of youth today the only true international community. More than this I think that we could agree on what kind of a world we would all want to build. It would be a world of independent nations moving toward international community, each of which protected and respected the basic human freedoms. It would be a world which demanded of each government that it accept its responsibility to insure social justice. It would be a world of constantly accelerating economic progress—not material welfare as an end in itself but as a means to liberate the capacity of every human being to pursue his talents and to pursue his hopes. It would, in short, be a world that we would be proud to have built.

Just to the north of here are lands of challenge and opportunity—rich in natural resources, land and minerals and people. Yet they are also lands confronted by the greatest odds—overwhelming ignorance, internal tensions and strife, and great obstacles of climate and geography. Many of these nations, as colonies, were oppressed and exploited. Yet they have not estranged themselves from the broad traditions of the West; they are hoping and gambling their progress and stability on the chance that we will meet our responsibilities to help them overcome their poverty.

In the world we would like to build, South Africa could play an outstanding role in that effort. This is without question a preeminent repository of the wealth and knowledge and skill of the continent . . .

But the help and the leadership of South Africa or the United States cannot be accepted if we—within our own countries or in our relations with others—deny individual integrity, human dignity, and the common humanity of man. If we would lead outside our borders, if we would help those who need our assistance, if we would meet our responsibilities to mankind, we must first, all of us, demolish the borders which history has erected between men within our own nations—barriers of race and religion, social class and ignorance.

Our answer is the world's hope; it is to rely on youth. The cruelties and obstacles of this swiftly changing planet will not yield to obsolete dogmas and outworn slogans. It cannot be moved by those who cling to a present which is already dying, who prefer the illusion of security to the excitement and danger which comes with even the most peaceful progress.

This world demands the qualities of youth; not a time of life but a state of mind, a temper of the will, a quality of the imagination, a predominance of courage over timidity, of the appetite for adventure over the love of ease. It is a revolutionary world we live in, and thus, as I have said in Latin America and Asia, in Europe and in the United States, it is young people who must take the lead. Thus you, and your young compatriots everywhere, have had thrust upon you a greater burden of responsibility than any generation that has ever lived.

"There is," said an Italian philosopher, "nothing more difficult to take in hand, more perilous to conduct, or more uncertain in its success than to take the lead in the introduction of a new order of things." Yet this is the measure of the task of your generation, and the road is strewn with many dangers.

First is the danger of futility: the belief there is nothing one man or one woman can do against the enormous array of the world's ills—against misery and ignorance, injustice and violence. Yet many of the world's greatest movements, of thought and action, have flowed from the work of a single man. A young monk began the Protestant Reformation, a young general extended an empire from Macedonia to the borders of the earth, and a young woman reclaimed the territory of France. It was a young Italian explorer who discovered the New World, and the thirty-two-year-old Thomas Jefferson who proclaimed that all men are created equal.

"Give me a place to stand," said Archimedes, "and I will move the world." These men moved the world, and so can we all. Few will have the greatness to bend history itself; but each of us can work to change a small portion of events, and in the total of all those acts will be written the history of this generation. Thousands of Peace Corps volunteers are making a difference in isolated villages and city slums in dozens of countries. Thousands of unknown men and women in Europe resisted the occupation of the Nazis and many died, but all added to the ultimate strength and freedom of their countries.

It is from numberless diverse acts of courage and belief that human history is shaped. Each time a man stands up for an ideal, or acts to improve the lot of others, or strikes out against injustice, he sends forth a tiny ripple of hope, and crossing each other from a million different centers of energy and daring those ripples build a current which can sweep down the mightiest walls of oppression and resistance.

"If Athens shall appear great to you," said Pericles, "consider then that her glories were purchased by valiant men, and by men who learned their duty." That is the source of all greatness in all societies, and it is the key to progress in our time.

The second danger is that of expediency; of those who say that hopes and beliefs must bend before immediate necessities. Of course, if we would act effectively we must deal with the world as it is. We must get things done.

But if there was one thing President Kennedy stood for that touched the most profound feelings of young people around the world, it was the belief that idealism, high aspirations, and deep convictions are not incompatible with the most practical and efficient of programs—that there is no basic inconsistency between ideals and realistic possibilities, no separation between the deepest desires of heart and of mind and the rational application of human effort to human problems. It is not realistic or hardheaded to solve problems and take action unguided by ultimate moral aims and values, although we all know some who claim that it is so. In my judgment, it is thoughtless folly. For it ignores the realities of human faith and of passion and of belief—forces ultimately more powerful than all of the calculations of our economists or of our generals.

Of course to adhere to standards, to idealism, to vision in the face of immediate dangers takes great courage and takes self-confidence. But we also know that only those who dare to fail greatly can ever achieve greatly.

It is this new idealism which is also, I believe, the common heritage of a generation which has learned that while efficiency can lead to the camps at Auschwitz or the streets of Budapest, only the ideals of humanity and love can climb the hills of the Acropolis.

A third danger is timidity. Few men are willing to brave the disapproval of their fellows, the censure of their colleagues, the wrath of their society. Moral courage is a rarer commodity than bravery in battle or great intelligence. Yet it is the one essential, vital quality of those who seek to change a world which yields most painfully to change. Aristotle tells us that "at the Olympic games it is not the finest and the strongest men who are crowned, but they who enter the lists . . . So too in the life of the honorable and the good it is they who act rightly who win the prize." I believe that in this generation those with the courage to enter the moral conflict will find themselves with companions in every corner of the world.

For the fortunate among us, the fourth danger is comfort, the temptation to follow the easy and familiar paths of personal ambition and financial success so grandly spread before those who have the privilege of education. But that is not the road history has marked out for us. There is a Chinese curse which says: "May he live in interesting times." Like it or not, we live in interesting times. They are times of danger and uncertainty; but they are also more open to the creative energy of men than any other time in history. And everyone here will ultimately be judged—will ultimately judge himself—on the effort he has contributed to building a new world society and the extent to which his ideals and goals have shaped that effort.

So we part, I to my country and you to remain. We are—if a man of forty can claim that privilege—fellow members of the world's largest younger generation. Each of us have our own work to do. I know at times you must feel very alone with your problems and difficulties. But I want to say how impressed I am with what you stand for and the effort you are making; and I say this not just for myself, but for men and women everywhere. And I hope you will often take heart from the knowledge that you are joined with fellow young people in every land, they struggling with their problems and you with yours, but all joined in a common purpose; that, like the young people of my own country and of every country I have visited, you are all in many ways more closely united to the brothers of your time than to the older generations of any of these nations; and that you are determined to build a better future. President Kennedy was speaking to the young people of America, but beyond them to young people everywhere, when he said that "the energy, the faith, the devotion which we bring to this endeavor will light our country and all who serve it—and the glow from that fire can truly light the world."

And, he added, "With a good conscience our only sure reward, with history the final judge of our deeds, let us go forth to lead the land we love, asking His blessing and His help but knowing that here on earth God's work must truly be our own."

The students interrupted the speech repeatedly with applause and, at the end, rose for a five-minute standing ovation. In his response, John Daniel, the vice president of NUSAS, told Kennedy, "We are proud to have had you speak to us and we are deeply ashamed of the fact that our government has not seen fit to welcome you to our country . . . You have given us a hope for the future. You have renewed our determination not to relax until liberty is restored, not only to our universities but to our land."

The Kennedys left the hall and walked through the thousands who had remained in the bitter cold of the South African winter night; many had waited in silence when it was discovered that the wires to the loudspeakers placed outside Jameson Hall had been cut. Later that evening, Ethel Kennedy visited Ian Robertson's flat, and she recounted for him in detail the moving scenes that his invitation had made possible.

Kennedy and his party spent the night at the Lanzerac Hotel in Stellenbosch, twenty-five miles east of Cape Town. On arrival, they learned that James Meredith, who had survived the sniping and riots when he integrated the University of Mississippi (with the help of Kennedy's Justice Department) four years earlier, had been shot in the back in an ambush alongside a highway near Hernando, Mississippi, as he led a 250-mile freedom walk from Memphis, Tennessee, to Jackson, Mississippi, on behalf of black voting rights.[12] No one knew if he would survive. The juxtaposition of the tremendous reaction Kennedy's visionary idealism had unleashed that evening in the supposed citadel of white supremacy, and the image of Meredith's body bleeding alongside a darkened country road in his native land, half a world away, tore at Kennedy. He was, traveling companion Walinsky remembered, "*really, really* . . . *terribly* upset."

Visiting the Cradle of Apartheid

UNIVERSITY OF STELLENBOSCH, SOUTH AFRICA

JUNE 7, 1966

Kennedy's University of Cape Town speech received extensive coverage around the world. The *New York Times* carried the story on its front page and included extensive excerpts of the speech; the other major New York papers also featured the address prominently, as did the newspapers in Washington, D.C., and Boston. Kennedy led the news in the London press for the second straight day as well, with the liberal papers applauding the speech as courageous and eloquent. Less impressed was the conservative *Daily Express,* which editorialized that "it is hard to see what useful purpose Senator Kennedy is achieving in South Africa."

Predictably, that question was receiving blanket coverage in South Africa itself. Johannesburg's *Rand Daily Mail* called Kennedy's Cape Town speech "one of the most stirring and memorable addresses the country has ever heard." The *Cape Times* provided eight pages of coverage, including reprinting the full text of the address. Its editorial, which noted Kennedy's "deep and patent sincerity," underlined for its readers the fundamental message of the speech:

> *What is profoundly significant for South Africa about this speech is that it is spoken in a world context, against a background of all the stirrings, aims, and pressures of the modern world. What the senator had to say to us he was also aware, as we must be aware, would be heard by the global community in which South Africans live and have to make their way—and with which they have to make their peace . . . It is suicidal in the long run to base a whole system of government on justification by human error and weakness in others as the South African government is inclined to do. The real drive in international politics and policies today is turning more and more to the positive realization of common human possibilities rather than purely negative defenses against human failures and inadequacies.*

The pro-government papers, however, gave the speech scant coverage and critical commentary.

On Tuesday, June 7, Kennedy spoke on the campus of the University of Stellenbosch, alma mater to all but one of South Africa's prime ministers, the first center of Afrikaner separatism, and the philosophical fountainhead of apartheid. His invitation had come from a student organization, but in the face of public criticism it had been rescinded; so instead, over a private lunch, Kennedy exchanged views with 280 student members of the Simonsberg men's residence hall.

The *Cape Times* reported that the controversy cast

> *a hesitant silence on the campus, as if the students did not quite know what to expect. But from the moment the youthful couple stepped from [tobacco baron Dr. Anton Rupert's] gleaming limousine, the ice was broken. Everyone within reach had his or her hand shaken—even the coloured cleaners and gardeners, several of whom ran off shouting with glee and waving aloft the "lucky" hand that had been shaken.*

Kennedy began his session with the Simonsberg men with brief prepared remarks.

If we—all of us—are to conquer anew the freedom for which our forebears gave so much, we must begin with a dialogue both full and free.

In the world of 1966 no nation is an island unto itself. Global systems of transportation and communications and economics have transformed our sense of geography and outmoded all the old concepts of self-sufficiency. Whether we wish it or not, a pattern of unity is woven into every aspect of the society of man.

We are protected from tetanus by the work of a Japanese scientist, and from typhoid by the work of a Russian. An Austrian taught us to transfuse blood, and an Italian to protect ourselves from malaria. An Indian and the grandson of a Negro slave taught us to achieve major social change without violence.

Our children are protected from diphtheria by the work of a Japanese and a German, from rabies by the work of a Frenchman, and cured of pellagra by the work of an Austrian. We all owe our very existence to the knowledge and talent and effort of those who have gone before us. We have a solemn obligation to repay that debt in the coin in which it was given: to work to meet our responsibilities to that greater part of mankind which needs our assistance, to the deprived and the downtrodden, the insulted and the injured. Those men who gave us so much did not ask whether we, their heirs, would be American or South African, white or black. And we must in the same way meet our obligations to all those who need our help, whatever their nationality or the color of their skin.

No longer can a spectator be certain that the blood and mud of the arena will not someday engulf him as well. No longer can any people be oblivious to the fate and future of any other. And no longer can any nation, no matter how wealthy or well armed, be as free as it once might have been to ignore a far-off war or warning, to shrug off another nation's crisis or criticism, or to defy the concerns or the contempt of mankind . . .

At times in our history, we have reacted too hastily and harshly to the fear of threats from within and without. But any times of suppression have been times of fear and stagnation, the years when the locust has eaten. We do not intend to repeat those years—even now, in the midst of a war in Vietnam. For we will not abolish the substance of freedom in order to save its shadow.

No nation would have so little confidence in the wisdom of its policies and its citizens that they dare not be tested in the free marketplace of ideas. Societies concerned with the importation of ideas are those which fear what Jefferson called "the disease of liberty." But those with confidence in their own future, in their citizens, and in the durability of their ideals, will welcome the exchange of views.

I am here in South Africa to listen as well as talk, less to lecture than to learn. Whatever our disagreements, neither your country nor mine is under any illusion that there is only one side to any issue or that either of us can coerce or quickly convert the other to share our point of view. But asserting disagreement without debates is as meaningless as asserting unanimity without discussion. Let us find out where we disagree, and why we disagree, and on what we can agree . . .

[Ours] is a world of change—unparalleled, unsettling, dizzying change. The certainties of yesterday are the doubts of today, and the folly and mockery of tomorrow. Every problem we solve only reveals a dozen more of increasing complexity.

Your country and mine have created wealth unmatched in the history of man; but we have not yet learned to turn that wealth to the service of all our people.

Your country and mine gained freedom from colonial domination and set an example for seventy nations around the world; but we have not yet learned how to help those new nations to achieve the economic, social, and political progress which their people demand and deserve . . .

In your country and mine, we fought for and achieved freedom for some of our people; but we have not yet learned, as Thomas Paine said, that "no man or country can be really free unless all men and all countries are free."

Kennedy's remarks were interrupted by the residence house's traditional student applause: the banging of soup spoons on the long wooden dining-hall tables. In the question-and-answer period that followed, Kennedy and the students found, through spirited dialogue, mutual appreciation and some significant agreement.

A number of the initial questioners defended apartheid as the only sensible response to whites' vast numerical inferiority; they argued that "separate development" would ultimately lead to separate nations, as India had been split to allow Hindu and Muslim autonomy. Kennedy quickly responded:

> *Shouldn't the colored man's voice be heard? The areas you have designated [for nonwhites] are strictly within the bounds you are going to give them. Are you really facing up to the situation in an honest, candid way? Somewhere in the future you are going to give them a separate country . . . and you are going to keep . . . 80 percent [of South Africa's land, including all of its arable portions] for the white people . . . We must face the fact that the white people are the minority group throughout the world.*

Particular press coverage was given to another exchange, stimulated by a student questioning whether Kennedy could make a worthwhile assessment of a country, especially one as complex as South Africa, on such a short visit.

"I don't say I am going to leave here as an expert on South Africa," Kennedy replied, "any more than I left South America an expert . . . But I have been here in good faith and have, as much as is possible under these rather difficult restrictions [he had previously referred to the government's harassment and failure to meet with him] . . . made a conscientious effort to know as much about South Africa as I can."

The students agreed, and at the end of the visit, the *Cape Times* reported that the "soup spoon tattoo . . . reached a thundering crescendo, lasting several minutes." The head student of the residence characterized the general opinion of his peers: that they were "impressed . . . [particularly because Kennedy] genuinely sought to understand a different point of view." Soon after, the Stellenbosch student body disaffiliated from the Afrikaner national student organization and inquired about joining NUSAS.

Such cracks in the facade of unified white domination were quickly noted by the government. When

Kennedy traveled to Durban that afternoon, one of the firebrand orators of the Nationalist Party, Deputy Minister Blair Coetzee, awaited him, scheduling an address to compete with Kennedy's visit to the University of the Natal that evening. At a party rally, Coetzee roared:

> *This little snip*[13] *thinks he can tell us what to do. He has only been in the country for three days and already he has the audacity to tell us what the remedy to our problems would be. There are a few true things in what he says. But we must show the rest of the world, including Mr. Kennedy, that the people of South Africa have never been more united . . . We will not be intimidated by America or England. We are growing militarily stronger every day and we could eat any other African state for breakfast. Kennedy can threaten as he likes, but we will show the world that our policy is the only one for South Africa.*

Kennedy, unaware of the content of Coetzee's attack but well aware of the viewpoint, addressed a university audience of ten thousand, the first on the trip to include many adults. Defenders of apartheid were again directly engaged; the aim, Kennedy said later, "was not simply to criticize but to engage in a dialogue to see if, together, we could elevate reason above prejudice and myth." Kennedy reminded the audience of the inherent absurdity of assuming that personal value was determined by skin color: "Maybe there is a black man outside this room who is brighter than anyone in this room—the chances are that there are many," he said to applause.

A questioner raised a standard justification for white supremacy: Rule by the black African majority, with a history of centuries of tribal warfare and no experience with self-government, could lead only to chaos and violence. Kennedy reminded them "that no race or people are without fault or cruelty. Was Stalin black? Was Hitler black? Who killed forty million people just twenty-five years ago? It wasn't black people, it was white."

Religion had provided another white rationalization, and a later questioner claimed biblical authority for black servitude. "But suppose God is black?" Kennedy replied. "What if we go to heaven and we, all our lives, have treated the Negro as inferior, and God is there, and we look up and He is not white? What then is our response?" Kennedy's questions, Ian Robertson remembered three years later, have "rung around South Africa ever since."

A Final Message to White South Africa

UNIVERSITY OF WITWATERSRAND, JOHANNESBURG

JUNE 8, 1966

Robert Kennedy's last full day in South Africa was a fitting climax to what the local press described as "an itinerary tight enough to leave the best-trained athlete exhausted." Kennedy began before dawn, taking advantage of a surprising reversal by the government, which had decided to grant his months-old request to see banned Chief Albert John Lutuli, spiritual leader of the country's blacks and the founder of the African National Congress. The Pretoria government had stripped Lutuli of his Zulu tribal leadership in 1952, and since that time (except when he traveled to receive the Nobel Peace Prize in 1961) he had been confined to his remote farm in a reserve area just outside of Stranger, some forty miles from Durban down the Valley of a Thousand Hills.

Kennedy brought a portable record player, and together he and the chief listened to President Kennedy's June 11, 1963, national address on civil rights, a subject that the president had called "a moral issue . . . as old as the scriptures and . . . as clear as the American Constitution." Then, to enjoy a degree of privacy impossible in Lutuli's house, where government security agents literally hung over their shoulders, the two took an extended walk in the surrounding fields.

"What are they doing to my countrymen, to my country?" Lutuli asked him. "Can't they see that men of all races can work together—and that the alternative is a terrible disaster for us all?" The chief left an indelible impression on his visitor; Kennedy praised Lutuli's "dignity, tolerance, and understanding" to a journalist upon returning to Johannesburg and later described him as one of the most impressive men he had ever met.

Under the terms of his banning, Lutuli could neither be quoted nor even referred to in print or in public, a stricture Kennedy promptly broke as he toured Soweto, a sprawling black township outside of Johannesburg. Kennedy brought his listeners the first news in nearly six years from their exiled leader.

While most of the area's adults were away at work, the children in the township turned out to greet the Kennedys, with hundreds surrounding them each time they stopped to shake hands or make impromptu speeches from the steps of a church, the roof of a car, or standing on a chair in the middle of a school playground. Thousands reached out to touch his hands, many crying, "Master, Master," despite Kennedy's repeated embarrassed requests, "Please don't use that word."

An accompanying South African journalist reported: "When the first group of children clustering thickly around the Kennedys outside the Regina Mundi Roman Catholic Church suddenly sang to them, Senator Kennedy and his wife fought back their tears. So did many others in the accompanying party of pressmen, officials and prominent citizens."

The residents of Soweto carefully clipped the newspaper photos and accounts of his visit and pasted them together as wallpaper, mementos they preserved for years afterward. The *Daily Rand*'s Jill Chisholm remembered, some years later, that "they were not a subservient people, the thousands who roared their approval, who rocked him on his roof, tugged at him, touched him in a display of belief and hope I'd never seen evoked before, never since."

Adulation by oppressed blacks was perhaps to be expected, but Kennedy's appeal was clearly broader. Adam Walinsky remembered that

> *the first couple of days [Kennedy's] only crowds were the students at the universities. But then, somewhere as the message started to get across . . . by the third and fourth day . . . there were just thousands of ordinary citizens . . . whites now, not blacks and Africans; not just Englishmen, but*

> *Afrikaners coming out there into the streets . . . to listen and . . . pulling at him and tugging at him and cheering him. It was really fantastic . . . what it turned into was a fantastic effort that I suppose if it'd lasted another couple of weeks would damn near have turned into an effort to overthrow the government. It may just have been the intoxication of the moment, but I tell you that by the fifth day there, we were convinced that we could, if that was what we wanted to do.*

It was a view echoed by local journalists, such as Anthony Delius, who covered the entire visit of the Kennedys and wrote that "the senator [had become] a kind of third political force in the country in less than a week."

In his final major address on June 8 at the University of Witwatersrand in Johannesburg on his last evening in South Africa, Kennedy (in Walinsky's words) "threw a lot of the restraints off . . . [in a speech] with a very different tone from the way he started, but by that time he could do it . . . it got much stronger."

Four days before, the *Rand Daily Mail* had reported an "overwhelming rush" on tickets for the twelve-hundred-seat Great Hall. On June 8, an enormous crush of people greeted Kennedy on his arrival.

Kennedy spoke to a crowd of seven thousand, including those listening via loudspeakers on the lawn outside the hall.

I have been in your country only a short time; yet you already have made a strong and deep impression . . . Everywhere I have been impressed with the warmth and the interest of all of the people of South Africa, of all political persuasions and races. Everywhere I have been impressed by your achievements, the wealth you have created in this continent which so sorely needs the blessings of progress.

Above all, I have been impressed with South African youth: not just those young in years, but those of every age who are young in a spirit of imagination and courage and an appetite for the adventure of life . . .

Will you sound the trumpet? And what is the battle, to which we all are summoned? It is the first battle for the future . . . The winds of freedom and progress and justice blow across the highest battlements, enter at every crevice, are carried by jet planes and communications satellites and by the very air we breathe.

Tomorrow's South Africa will be different from today's, just as tomorrow's America will be different from the country I left these few short days ago: different for the astronauts who returned from their journey—and for James Meredith, who did not complete his journey. Our choice is not whether change will come but whether we can guide that change in the service of our ideals and toward a social order shaped to the needs of all our people. In the long run we can master change, not through force or fear, but only through the free work of an understanding mind, through an openness to new knowledge and fresh outlooks, which can only strengthen the most fragile, and the most powerful human gifts: the gift of reason.

Thus those who cut themselves off from ideas and clashing convictions not only display fear and enormous uncertainty about the strength of their own views; they also guarantee that when change comes, it will not be to their liking. And they encourage the forces of violence and passion which are the only alternatives to reason and the acts of minds freely open to the demands of justice.

Justice—a demand which has echoed down through all the ages of man. This is the second battle to which we are summoned. And let no man think that he fights this battle for others; he fights for himself, and so do we all. The Golden Rule is not sentimentality but the deepest practical wisdom.

For the teaching of our time is that cruelty is contagious, and its disease knows no bounds of race or nation. Where men can be deprived because their skin is black, in the fullness of time others will be deprived because their skin is white. If men can suffer because they hold one belief, then others may suffer for the holding of other beliefs. Freedom is not money, that I could enlarge mine by taking yours. Our liberty can grow only when the liberties of all our fellow men are secure; and he who would enslave others ends only by chaining himself, for chains have two ends, and he who holds the chain is as securely bound as he whom it holds. And as President Kennedy said at the Berlin Wall in 1963, "Freedom is indivisible, and when one man is enslaved, all are not free." . . .

There are those who say that the game is not worth the candle—that Africa is too primitive to develop, that its peoples are not ready for freedom and self-government. But those who say these things should look to the history of every part and parcel of the human race. It was not the black man of Africa who invented and used poison gas or the atomic bomb, who sent six million men and women and children to the gas ovens and used their bodies as fertilizer . . . And it was not the black men of Africa who bombed and obliterated Rotterdam and Shanghai and Dresden and Hiroshima.

We all struggle to transcend the cruelties and the

follies of mankind. That struggle will not be won by standing aloof and pointing a finger; it will be won by action, by men who commit their every resource of mind and body to the education and improvement and help of their fellow man.

And this is the third aspect of our battle: to fight for ourselves as individuals, and for the individuality of all.

We are patriots. We believe in our countries and wish to see them flourish. But the countries we love are not abstractions . . .

History is full of peoples who have discovered it is easier to fight than think, easier to have enemies and friends selected by authority than to make their own painful choices, easier to follow blindly than to lead, even if that leadership must be the private choice of a single man alone with a free and skeptical mind. But in the final telling it is that leadership, the impregnable skepticism of the free spirit, untouchable by guns or police, which feeds the whirlwind of change and hope and progress in every land and time.

Again, the questioning at the end of the session included the now standard give-and-take about the philosophical and political aspects of apartheid, and about the relationship between South Africa and the United

States. Kennedy wrote later that "the final question was the most difficult." A student asked, "How do you keep up a dialogue with people who change the rules of debate to suit themselves, tell you what you can talk about, insist upon being judge and referee, and ban you when you say too much for their liking?" Kennedy replied, "The only alternative is to give up, to admit that you are beaten. I don't know about that. I have never admitted that I am beaten." And, as he later wrote in a 1966 *Look* magazine article titled "Suppose God Is Black," in his "judgment, the spirit of decency and courage in South Africa will not surrender."

That spirit was admirably reflected by Martin Shule, a student selected to deliver a formal response to Kennedy's address. Shule told his peers: "We must now cast off all self-protective timidity, and we must now willfully and deliberately descend into the arena of danger to preserve the independence of thought and conscience and action which is our civilized heritage. We must now set ourselves against an unjustifiable social order and strive energetically and selflessly for its reform."

Five and a half hours after his day in Johannesburg had ended, and only five days after his arrival in the country, Kennedy left for Nairobi, Kenya, en route to Ethiopia and East Africa. In South Africa, memories of his visit continued to monopolize press commentary and public conversation. "Even in remote areas such as South West Africa," the United States embassy in Pretoria reported back to Washington, "the Kennedy speeches were passed from hand to hand in the African townships."

Most Afrikaans papers were scathing. *Die Vaderland* complained that far too much attention had been paid to the foreigner, and "his visit has not made us wiser." *Die Burger* took Kennedy to task for indulging in "fiery verbal idealism" and "idealistic politicking," which the paper claimed was an escapist dodge of someone who should more profitably have remained home to focus on "shotguns in Mississippi or burning monks in Vietnam."

The English-language newspapers saw Kennedy's impact differently. Writing in the *Cape Argus*, commentator H. G. Lawrence noted that "in the course of four days this dynamic young American was able to enthuse vast crowds and bring a strong breath of fresh air into this laager-ridden country." The *Cape Times* reprinted Frank Taylor's piece for the London *Daily Telegraph*, concluding that

> *for people weaned on the stolid, rather plodding politics of Afrikanerdom, the tactics of the*

> *barnstorming junior senator from New York had a stimulating but somewhat frightening effect . . . If [his effect] . . . on adult South Africa was profound, his influence on the young students who invited him to the Republic was nothing short of astonishing. But the senator has now left and it is hoped that those students who rode the crest of an unaccustomed wave of frankness and self-expression with him for five heady days have not left themselves open to official pressure or intimidation.*

Anthony Delius of the *Cape Times* was more hopeful: Kennedy's insistence "that he and all South Africans . . . belonged to the same world . . . came as a sign to [the country's youth] that they were not cut off from their generation—and, by using their abundant energy and imagination, need never be."

The *Rand Daily Mail* released a booklet after the visit, with numerous pictures, the complete text of each of the formal speeches, and some contemporaneous commentary. It concluded with a reprint of the paper's editorial from the day of Kennedy's departure, which ended: "Thank you a thousand times for what you have done for us. Come back again. You have a place in our hearts."

Kennedy hoped to come back. His old nemesis, ex-Minister Louw, told the *Cape Times* the day after Kennedy left that the government would never grant him a return visa. Three days later, the nationalist Sunday newspaper *Beeld* confirmed that Kennedy could not return. As its sister paper *Dagbreek,* chaired by the president of South Africa, said, the government "had enough of Kennedy's antics" and vowed that vigilance over "subversive elements" within the country would increase in the wake of the visit.

On returning to the United States, Kennedy wrote to all American companies with operations in South Africa, asking them to take all possible steps to improve the lot of their black employees. He felt such direct measures were far preferable to the principal leverage being touted in the United Nations at the time: As Kennedy told a press conference at the airport on arrival in New York, he opposed trade sanctions against South Africa, believing that such measures would most severely hit the country's blacks.

But South Africa was not the lead item in the minds of the reporters gathered at the airport. That day Senator Wayne Morse, a Democratic antiwar ally of Kennedy's, vowed to oppose the renomination of Lyndon Johnson and to urge Kennedy to run. Once again, political jockeying took press precedence, and the headlines went to Kennedy's announcement that he had no intention of running and anticipated supporting Johnson.

Whatever window Kennedy's visit to tortured South Africa threw open was soon closed, and the generation of students who had cheered him so wildly grew to middle age under the same system of apartheid they had sworn to end. It took more than two decades for legal discrimination to be ended, and nearly three decades for South Africa to move toward peace and equal justice. But Kennedy's exhortations were not in vain. In early 1992, two years after Nelson Mandela left Robben Island, a delegation from the African National Congress visited the John F. Kennedy Presidential Library in Boston. When chief archivist Will Johnson asked his guests whether any of them remembered Robert Kennedy's five days in their country, twenty-four years before, he was stunned as five freedom fighters in their forties, all alumni of South African jails and oppression, rose and recited from memory passages from the Day of Affirmation address. Yes, we remember, they told Johnson; it kept us strong through all the years of darkness.

The Goals of American Foreign Policy: Columbus Day Dinner

WALDORF ASTORIA HOTEL; NEW YORK, NEW YORK

OCTOBER 11, 1966

Kennedy's travels in South Africa and Latin America were vital to his views of a rapidly changing world community and of America's role in it—views seasoned by visits to twenty countries during the three years of his brother's administration and, before that, equally intensive trips in his youth throughout Europe, to the Middle East, around Asia, and in the Soviet Union and Eastern Europe. To an extraordinary degree, he utilized the best that American higher education and family wealth could offer and allied them with his experiencing nature—that strain of personal character that led him to avoid the staged embassy tours, to venture far from the ordinary haunts, and always to probe for the deeper background and authentic nuance that were keys to a defining vision of a nation and its people.

The integration of disparate cultures into a world with lasting peace, and the strategies that would allow America to shape that goal, validate our revolutionary ideals, and protect our national interests can be seen throughout many of Robert Kennedy's addresses. A notable example was given to a Columbus Day dinner in New York City on October 11, 1966.

The year 1492 symbolizes the three parts of the American experience: our past conquests, our present dangers, and the most spacious of our future hopes.

At the same moment that men were flaunting danger to master a primitive continent, the civilization of the world was reaching one of its greatest moments. In the Italy from which Columbus came, Lorenzo the Magnificent was ending his rule, Bellini and Michelangelo were creating wonders of beauty, and Leonardo da Vinci was penetrating the mysteries of science and art. That Italian Renaissance was destined to shape the thought and history of our world as profoundly as the voyages of Columbus.

We are as much the inheritors of one as of the other. The first: daring, the building of new worlds, the search for opportunity and even for wealth and power. The second: the desire to build a civilization to liberate the full capacities of its own people and to take upon itself the burden of helping to illuminate the lives of all men.

For the first time in five centuries since Columbus, we have the chance, the power, and the obligation not only to build a great nation for ourselves but to be powerful, shaping forces in that world from which he came, and in obscure continents he barely knew.

For most of those five centuries, the rest of the world mattered to us only when it touched our own concerns. During most of our history as a nation we tried to avoid entanglement in the affairs of others. Yet today, our armies fight in Asia and guard the borders of Europe.

. . . Moreover, our achievements are reshaping the planet . . . Our weapons have changed large-scale warfare from national policy into a threat to the existence of the human race . . . and the words "give me liberty or give me death" have been scrawled on sidewalks and walls from Indonesia to Nigeria.

Wealth and power and influence are thus inescapable facts. For much of the postwar period, we have used that strength to meet threat after threat: deterring the ambitions of conquerors, rebuilding Europe, organizing our allies, seeking guarantees of peace, and trying to resolve conflicts . . .

All of these threats and conflicts are still with us . . . We have grown accustomed enough to our danger and responsibility so that we can now ask ourselves the question: For what reason? To what larger purpose must we put our might and energy and the fantastically varied skills of our people?

The answer begins at the heart of our concern as a nation: the well-being, freedom, and security of our own people. Those, we must be willing to defend with all our resources and, when necessary, with our lives. But that alone is not enough. Pericles said of Athens: "The admiration of the present and succeeding ages will be ours, since we have not left our power without witness, but have shown it by mighty proofs."

Those proofs are not the forgotten conquests of the Athenian fleet; just as the proofs of fifteenth-century Italy are not the armed strength and wars of her cities. Through the history of the world, the boundaries of great empires have faded and dissolved, their cities fallen into decay, and their wealth scattered.

What remains is what they accomplished of enduring value and what they stood for. What remains is the contribution they made to the unity and knowledge and understanding of man. What remains is what they added to the hopes and well-being of human civilization and to its capacity for future progress.

None of us here, as individuals, seek success or wealth purely for its own sake. We all hope to make a larger contribution: to our families, our occupation or profession, our community, and our country. This must be true also for America as a nation, if we wish it to take that luminous and lasting place in history which is now within its grasp.

Our country began as a center of hope, not only for those who came here but for those who did not. Thomas Jefferson told us, "We are pointing the way to struggling nations who wish, like us, to emerge from their tyrannies also." . . .

Think how our world would look to a visitor from another planet as he crossed the continents. He would find great cities and knowledge able to create enormous abundance from the materials of nature. He would witness exploration into understanding of the entire physical universe, from the particles of the atom to the secrets of life. He would see billions of people separated by only a few hours of flight, communicating with the speed of light, sharing a common dependence on a thin layer of soil and a covering of air. Yet he would also observe that most of mankind was living in misery and hunger, that some of the inhabitants of this tiny, crowded globe were killing others, and that a few patches of land were pointing huge instruments of death and war at others. Since what he was seeing proved our intelligence, he could only wonder at our sanity.

It is this monstrous absurdity—that in the midst of

such possibility, men should hate and kill and oppress one another—that must be the target of the modern American Revolution.

This does not mean that we neglect our own interest. For this is our interest. That interest is not just to prevent our own devastation or to find markets for our goods. In a dwindling world where a century of change tumbles into a decade, pursuit of goals so narrowly conceived would bring disaster.

An America piled high with gold, and clothed in impenetrable armor, yet living among desperate and poor nations in a chaotic world, could neither guarantee its own security nor pursue the dream of civilization devoted to the fulfillment of man.

Our true interest, therefore, is to help create a world order to replace and improve that shattered when World War I opened the doors to the twentieth century; not an order founded simply upon balance of power or balance of trade; but one based on the conviction that we will be able to shape our own destiny only when we live among others whose own expectations are unscarred by hopelessness, or fear of the strong, or the ambition to master other men . . .

The spirit and meaning of this nation will be in jeopardy if we grapple it to ourselves rather than extend it to others. History is a relentless master. It has no present, only the past rushing into the future. To try to hold fast is to be swept aside.

I cannot present nor would it be appropriate to present a detailed guide for working out the ideals of American purpose. Yet certain objectives are becoming clearer.

It is not permissible to allow most of mankind to live in poverty, stricken by disease, threatened by hunger, and doomed to an early death after a life of painful labor. It is easy to debate, at great skillful length, the merits of one particular form of foreign assistance or another, the need for self-help, or the mechanics of development. The fact is that the fortunate fraction of mankind now has the technology and the knowledge to improve all these afflictions, and we must seek huge leaps of imagination and effort to shatter the frustrating and resistant barriers between human capacity and human need. We must also help find a way to dissolve the attitudes which permit men to indulge those passions and ambitions which keep the world in constant conflict, and which threaten the survival of all of us . . .

Nor is the peace we seek simply the absence of armed conflict or hostile division. It is the creation among nations of a web of unity, woven from the strands of economic interdependence, political cooperation, and a mounting flow of people and ideas . . .

We must, if we seek not merely to lead but to lead greatly, act consistently with our belief in human freedom and equality. Those are the seminal values of our entire history. We realize that for many, liberty today is often a remote pursuit, lacking urgency to those enslaved by material want. Nor should we, if we could, compel other countries to adopt our principles. Yet there should be no doubt that we stand—in Africa or Asia or in Latin America, and in the United States itself—on the side of equality and increasing freedom; never yielding that position to the demands of temporary expediency or short-run realism. For if we allow immediate considerations, one by one, to chip away proclaimed ideals and values, then we soon stand for nothing at all, except ourselves . . .

Nor should our own success and good fortune allow us to treat with scorn or condescension or lack of understanding those who are struggling to build nations under the crushing weight of instability, desperate economies, and fragile society . . .

All of these are guidelines which must, of course, be translated into policies and acts in every part of the world over many years. Yet they may help us toward an opportunity which is granted to few generations in few lands. For we are a nation which reached the height of its power and influence at a time when the old order of things is crumbling and a new world is painfully struggling to take shape.

It is a moment as fully charged with opportunity as that granted to Columbus or the heroes of the Italian Renaissance. It offers to this nation the chance for great achievement—or perhaps the greatest and most destructive of failures. It is a voyage more hazardous and uncertain than that which we celebrate today. For we seek to cross the dark and storm-scarred seas of human passion and unreason, ignorance, and anger. They were as uncharted in Columbus's time as they are today. Yet we have been thrust out upon them by our mastery of the continent he discovered and the knowledge his age began. The way is uncertain and the trip is charged with hazard. Yet perhaps we can say, in the words of Garibaldi to his followers: "I do not promise you ease. I do not promise you comfort. But I do promise you these: hardship, weariness, and suffering. And with them, I promise you victory."

America in Vietnam

The Scylla and Charybdis of American policy in the 1960s, the twin shoals that threatened to tear apart the ship of state, were race relations and the war in Vietnam. The history of Vietnam, even the limited period of America's involvement there during Robert Kennedy's adulthood, is rich and complex. Culminating in an undeclared war that challenged three administrations, took the lives of nearly sixty thousand Americans, and wounded more than five times that number—with Vietnamese civilian and military casualties in the millions—the war in Vietnam continued to shape military and diplomatic decision-making in the United States and abroad more than forty years after the last American service member left the country.[14]

During the years Robert Kennedy served in the

Senate, majority popular opinion continued a gradual change regarding racial matters; but concerning the moral underpinnings and likely outcome of the war, the shift in course was marked. Kennedy's evolution, best analyzed through his speeches, was pronounced and generally came well in advance of the public mood.

Robert Kennedy visited Vietnam with his brother a year before John first ran for the United States Senate. In 1951, a critical year in Vietnam's post–World War II history, forces that were unleashed with the Japanese surrender collided with increasing intensity.[15] As Japanese troops withdrew in August 1945, the French reasserted their dominance over their former colony. The Vietminh, led by Ho Chi Minh, countered by declaring the country independent on September 2, 1945, with a statement that began by repeating Jefferson's introduction to the United States Declaration of Independence. Ho initially hoped that America (with whom he had allied in the struggle against Vietnam's Japanese occupiers) would support national independence against French recolonization.

American and British governments feared that the Vietminh were hardline Communists and accordingly gave tacit approval for French troops to intervene. During the three years following the Kennedy brothers' 1951 visit to Vietnam, American military aid to the French effort amounted to another $1.8 billion. The French proved militarily incapable of controlling the country and were soon defeated, their forces withdrawing shortly after the surrender of their fortress at Dien Bien Phu in May 1954. Because President Eisenhower doubted the long-range military prospects in perpetuating a government devoid of popular support, he chose not to commit the United States to direct military involvement.

When John Kennedy was inaugurated as president in January 1961, he and his brother regarded neighboring Laos, rather than Vietnam, as far more critical to United States interests. Although he had felt initially that military intervention might be required to repel Communist advances in Laos, the fiasco of a United States–backed invasion attempt in Cuba (the Bay of Pigs) later in that year convinced the president to seek a negotiated solution in Laos, and an accord was signed in Geneva in July 1962.

Subsequently the Kennedy administration's attention in Indochina focused on Vietnam. Ngo Dinh Diem was South Vietnam's first premier after the 1954 Geneva Accords ended the war between the Vietminh (from North Vietnam) and the French. Fielding nearly a quarter of a million army, guard, and related forces, Diem was locked in an inconclusive effort against an estimated twelve thousand Vietcong (the Vietminh's southern military arm). In the United States, powerful domestic forces (including the *New York Times* and the Pentagon) argued for immediate, direct American support to ensure Diem's survival.

President Kennedy agreed to commit one hundred military advisers and four hundred Green Berets (special forces troops). This was far less than the Pentagon had argued was the minimum required, but the president expressed great discomfort with a military approach in Vietnam. In this he was bolstered by General Douglas MacArthur (the hero of the Pacific in World War II and head of Allied forces in Korea, until he was removed by President Truman for exceeding his authority), who warned that American military intervention would be a mistake.

The principal alternative considered by the Kennedy administration was a strategy of counterinsurgency, which arose from a premise that the struggle was predominantly political and economic, and therefore that substantial efforts had to be placed on land reform, local development, and similar efforts. Militarily, counterinsurgency called for use of the guerrilla tactics employed by the Vietcong, rather than reliance on the massive air power and overwhelming ground-troop strength favored by the Pentagon. Robert Kennedy is credited with having coined the term *counterinsurgency,* and he became its principal exponent in the administration. Michael Forrestal, the president's assistant for Far Eastern affairs, recalled that Robert Kennedy believed the strategy would have particular value in Latin America, but

> *he would have included Southeast Asia . . . Bobby was spending a lot of his energy on the Special Group CI [for counterinsurgency], which was a committee in the White House that met once a week . . . a very high-level group. It was Bobby's major touch with foreign affairs. He wanted to be sure that the government was prepared for counterinsurgency efforts. These never involved the thought of American troops. Counterinsurgency, in Bobby's mind, was largely a civilian effort backed up by the right kind of supplies delivered at the right time to the right place.*

President Kennedy was more deeply involved than his brother in the evolution of United States policy in Vietnam, but even for the president, it was a matter far less significant than a host of other foreign and domestic issues. In July 1962, President Kennedy instructed Defense Secretary Robert McNamara to prepare a phased withdrawal of American military personnel, moving toward complete disengagement by the end of 1965. He made it repeatedly clear, in public statements and in private, that he opposed dispatching American ground troops to Vietnam and that he felt the Constitution required congressional concurrence if such military involvement were to be instituted. In the beginning of November 1963, a cabal of South Vietnamese generals (confident of United States support of their efforts) toppled and soon murdered Diem.

Robert Kennedy attended all of the principal White House meetings through the late summer and early fall of 1963, the months preceding the coup. At a particularly critical National Security Council meeting on September 6, 1963, he urged that the immediate question of Diem's future be considered in the context of America's stated goal: helping the Vietnamese people resist a Communist takeover. If Diem's regime was incapable of doing so, but the generals were, then (Kennedy believed) our responsibility to the Vietnamese dictated that we back the coup. But, Kennedy continued, if no conceivable leadership in the South could defeat the Vietcong, then the United States should promptly disengage. Still, John Kennedy was less clear in public about the extent of America's commitment to a substantial presence in Vietnam and to the Diem regime. In part, he did not want the Republicans to campaign against him in 1964 on the premise that Kennedy was losing Vietnam to the Communists. Thus he repeatedly asserted (in the face of press questioning) that he had no intention of abandoning Vietnam.

Michael Forrestal remembered that in the months preceding John Kennedy's assassination, Robert expressed increasing doubts about the entire U.S. effort

> *not from a moral point of view, but from a pragmatic point of view . . . Did the United States have the resources, the men, and the philosophy, and the thinking to have anything useful to contribute . . . in a country as politically unstable as South Vietnam was? . . . He raised questions—hard questions—which most of us had assumed had been answered many years before. He began forcing people to take a harder look at what we were doing there, and whether or not we were really capable of doing it.*[16]

An Initial Warning Against a Widening War

UNITED STATES SENATE

MAY 6, 1965

Although he continued as a cabinet member for nearly a year under Lyndon Johnson, Robert Kennedy ceased active involvement in policy regarding Vietnam after his brother's death. His views during 1964 were seldom expressed publicly and only in general terms. In a commencement address at the California Institute of Technology on June 8, 1964, he said:

> *To the extent that guerrilla warfare and terrorism arise from the conditions of a desperate people, we know that they cannot be put down by force alone. The people themselves must have some hope for the future. There must be a realistic basis for faith in some alternative to Communism . . . Far too often, for narrow tactical reasons, this country has associated itself with tyrannical and unpopular regimes that had no following and no future. Over the past twenty years we have paid dearly because of support given to colonial rulers, cruel dictators, or ruling cliques void of social purpose . . . Ultimately, Communism must be defeated by progressive political programs which*

wipe out the poverty, misery, and discontent on which it thrives.

Vietnam played no part in Kennedy's subsequent campaign for the Senate. The same month he was sworn into office, however, he heard from sources in the administration that Johnson was leaning toward abandoning the United States presence in Vietnam. Yet despite his concerns about American engagement, Kennedy believed that withdrawal would betray legitimate U.S. interests—and the Vietnamese who had supported us. He wanted to see a recommitment to counterinsurgency. He instructed Ed Guthman and Roger Hilsman (his brother's chief State Department expert on Vietnam) to draft a speech opposing both withdrawal and a conventional military strategy.

The speech was not completely written when the Vietcong launched a heavy attack on American barracks at Pleiku on February 7, 1965. Johnson responded by announcing bombing raids on North Vietnam and sending two hundred thousand U.S. troops to the South. Kennedy went to see the president in April, urging a bombing pause to test the Vietcong's interest in a negotiated settlement.

Instead, Johnson forwarded to Congress his request for $700 million in supplemental appropriations for the massive troop deployment and continued bombing. Days before, United States troops had rolled through the capital of the Dominican Republic. Kennedy's initial instinct was to oppose the appropriation as an unnecessary step in the service of an unwise policy. He consulted with Arthur Schlesinger Jr. and Burke Marshall, who both urged the senator to vote in favor of the appropriations but to state in detail his concerns and the limitations of his support.

With some modifications, Kennedy decided to deliver the remarks drafted by Schlesinger and approved by Marshall, encompassing events in both the Caribbean and Vietnam. It marked the first time Robert Kennedy spoke on the Senate floor.

Mr. President, I vote for this resolution because our fighting forces in Vietnam and elsewhere deserve the unstinting support of the American government and the American people. I do so in the understanding that, as Senator [John] Stennis [Democrat, Mississippi] said yesterday: "It is not a blank check . . . If [the president] substantially enlarges or changes [his present policy], I would assume he would come back to us . . ."

We confront three possible courses in Vietnam. The

first is the course of withdrawal. Such a course would involve a repudiation of commitments undertaken and confirmed by three administrations. It would imply an acquiescence in Communist domination of south Asia—a domination unacceptable to the peoples of the area struggling to control and master their own destiny. It would be an explicit and gross betrayal of those in Vietnam who have been encouraged by our support to oppose the spread of Communism. It would promote an inexorable tendency in every capital to rush to [Beijing] and make the best possible bargain for themselves. It would gravely, perhaps irreparably, weaken the democratic position in Asia.

The second is the course of purposely enlarging the war. Let us not deceive ourselves; this would be a deep and terrible decision. We cannot hope to win a victory over Hanoi by such remote and antiseptic means as sending bombers off aircraft carriers . . .

The course of enlarging the war would mean the commitment to Vietnam of hundreds of thousands of American troops. It would tie our forces down in a terrain far more difficult than that of Korea, with lines of communication and supply far longer and more vulnerable . . .

Both of these courses—withdrawal and enlargement—are contrary to the interests of the United States and to humanity's hope for peace. There remains a third course—and this, I take it, is the policy of the administration, the policy we are endorsing today. This is the course of honorable negotiation . . .

To create the atmosphere for negotiation in these conditions, we must show Hanoi that it cannot win the war, and that we are determined to meet our commitments no matter how difficult. This is the reason and the necessity, as I understand it, for the military action of our government. But I believe we should continue to make clear to Hanoi, to the world, and to our own people that we are interested in discussions for settlement. I believe that our efforts for peace should continue with the same intensity as our efforts in the military field. I believe that we have erred for some time in regarding Vietnam as purely a military problem when in its essential aspects it is also a political and diplomatic problem. I would wish, for example, that the request for appropriations today had made provision for programs to better the lives of the people of South Vietnam. For success will depend not only on protecting the people from aggression but on giving them the hope of a better life which alone can fortify them for the labor and sacrifice ahead.

The Moral Implications of the War Effort

FACE THE NATION

NOVEMBER 26, 1967

Over the course of the next two and a half years, America's commitment of troops and treasure increased dramatically, yet despite repeated pronouncements from the White House and the field that victory was imminent, U.S. casualties escalated—in 1966, exceeding those suffered by the South Vietnamese military. Bombing of North Vietnam increased, with no discernible diminution of the fighting in the south. Politically, most American voters still supported the war effort, and President Johnson and the media scrutinized every Kennedy action on Vietnam as carefully calculated to enhance his presumed run against the incumbent in 1968.

For his part, Kennedy tried to calibrate his statements or silences in ways that would influence the president or the public toward what Kennedy came to regard as the only path to peace. But it became inescapably clear that Johnson's personal animosity and political survival instincts would never lead to adoption of any Kennedy proposal—and was more guaranteed to boomerang.[17] Finally, on March 2, 1967, Kennedy broke with the administration. He shouldered his share of past decisions: "Three presidents have taken action in Vietnam. As one who was involved in many of those decisions, I can testify that if fault is to be found or responsibility assessed, there is enough to go around for all—including myself." Kennedy then detailed the horrors the war had inflicted on civilians and combatants, and he proposed a plan for a negotiated settlement.

Less than two weeks later Kennedy visited the University of Oklahoma and was questioned about his stand on student deferments from the draft. Kennedy replied that he had the money to allow his children to attend college but that it was a clear moral wrong for the future of America's sons to rest on the size of their parents' pocketbooks. Oklahoma's Democratic Senator Fred Harris, who accompanied Kennedy, recalled that "there was hissing and booing."

Kennedy then said, "Let me ask you a few questions." How many favor continuing deferments? Overwhelming cheers. How many support increasing the military effort? Again, predominant approval. "Let me ask you one other question . . . How many of you who voted for the escalation of the war also voted for the exemption of students from the draft?" There was, Harris said, a "giant gasp," followed by silence and then thunderous applause. No one had confronted them like this before, and they had the honesty to appreciate the moral mirror. The national view was different: A May Gallup poll recorded that half of the students surveyed described themselves as hawks on the war, with only 35 percent calling themselves doves.

As 1967 drew to a close, more than half a million U.S. troops were stationed in Vietnam, and North Vietnam suffered heavier bombing than that visited on all fronts during World War II. Protests against the war became larger and angrier. In October, Kennedy bested Johnson in a Harris poll, 52 to 32 percent. Also that fall, Kennedy's book *To Seek a Newer World* was published; more than one third of the book was devoted to a discussion of the war and Kennedy's program for peace.

But while his views on the war pushed Kennedy further into open confrontation with the president, his political training (and the more experienced of his advisers) continued to argue against challenging Johnson. While peace advocates urged Kennedy to run, he agonized over a Hobson's choice of seeking the nomination—and appearing nakedly ambitious, risking splitting the party, and probably losing to the incumbent (despite Kennedy's temporary lead in the polls)—or supporting Johnson, and looking hypocritical, in light of his stated abhorrence of the war and concern for the administration's abandonment of the nation's cities. On Sunday, November 26, 1967, Kennedy appeared on *Face the Nation,* throwing both his political indecision and passionate war opposition into high relief.

I have found it, over the period of the last eighteen months particularly, very difficult to talk about some of these matters without getting involved in personalities. When I criticized the war in Vietnam in a major speech back in February 1966, after the initial stories, it was placed purely . . . on the basis of a personality struggle between President Johnson and myself . . . that I would like to still have President Kennedy president of the United States, a resentment of Johnson as president of the United States. It has never really been analyzed on the basis of my criticism . . . I think that having it a personality struggle rather than an issue question has been damaging to the country as a whole and damaging, really, to the consideration of these matters. So I think that . . . if I ran for president . . . I would not strengthen the . . . dialogue that is taking place in connection with these issues, but in fact I would weaken it. It would immediately become a personality struggle between me, as an overly ambitious figure trying to take the nomination away from

President Johnson, who deserves it because of the fact that he is not only president but served the Democratic Party and the country as president for four or five years . . . I am going to continue to talk about these issues. But I am not talking about these issues as a competitor to anybody, I am talking about these issues as issues which I think are important to the American people . . . I think that what we are doing at the present time in Vietnam is a mistake. I think that the course that we are following is an error. But I am saying that as a United States senator, and I want to have what I say analyzed on that basis . . .

I do not think that to dissent here in the United States or those who disagree should be confined to those who are young. I think that when we sign up for the Democratic Party we don't say that we are never going to disagree. I think that there is much to disagree about . . .

First we were making the effort there so that people would have their own right to decide their own future and could select their own form of government, and it wasn't going to be imposed on them by the North Vietnamese, and we had the support of the people of South Vietnam. I think that is why we were involved in that struggle. That is certainly the way I looked at it when I was in President Kennedy's administration and when

I was with President Johnson. Now we turned, when we found that the South Vietnamese haven't given the support and are not making the effort; now we are saying we are going to fight there so that we don't have to fight in Thailand, so that we don't have to fight on the West Coast of the United States, so that they won't move across the Rockies.

But do we—our whole moral position, it seems to me, changes tremendously. One, we're in there, we're helping people. We're working with them. We're fighting for their independence. Second, we're—and we're killing the enemy and we're also killing many civilians, but we're doing it because they want it. Now we've changed and switched. Maybe they don't want it but we want it. So we're going in there and we're killing South Vietnamese; we're killing children; we're killing women; we're killing innocent people because we don't want to have the war fought on American soil, or because they're twelve thousand miles away and they might get to be eleven thousand miles away.

Our whole moral position changes, it seems to me, tremendously. Do we have the right here in the United States to perform these acts because we want to protect ourselves, so that . . . it is not a greater problem for us here in the United States? I very seriously question whether we have that right . . . I think other people are fighting it, other people are carrying the burden . . . but this is also our war. Those of us who stay here in the United States, we must feel it when we use napalm, when a village is destroyed and civilians are killed. This is also our responsibility. This is a moral obligation and a moral responsibility for us here in the United States.

And I think we have forgotten about that. And when we switched from one point of view to another, I think that we have forgotten about it. And I think that it should be discussed and all of us should examine our own conscience of what we are doing in South Vietnam. It is not just the fact that we are killing North Vietnamese soldiers or Vietcong; we are also responsible for tens and tens of thousands of innocent civilian casualties, and I think we are going to have a difficult time explaining this to ourselves . . .

All acts of aggression, death, and destruction occur in war—sometimes wars are essential, are necessary, are going to occur. But we should also consider the price that we are paying. It is not just the Americans that are being killed, the Americans that are being wounded, and the price that we are paying so that we can't do the kinds of things that we should, but we have a moral position around the world . . . We can't lose that, as it appears to be that we are doing in Vietnam.

Why can't the president of the United States or the vice president of the United States travel freely around the world anymore? It is because of Vietnam. Why can't they go through Latin America? Why can't they travel through Europe? Why can't they even travel freely throughout our own country? . . .

We should look at it in an objective way, at what we are doing, what we are trying to do, and what this country stands for, both . . . internally within the United States and what we have to stand for around the rest of the globe. If this country is going to mean anything . . . we love our country but we love our country for what it can be and for the justice it stands for and what we are going to mean to the next generation. It is not just the land, it is not just the mountains . . . And that is what I think is being seriously undermined in Vietnam, and the effect of it has to be felt by our people.

No cold transcript can convey Kennedy's emotionally charged comments about the horror of the war. Walinsky, at the taping, remembered watching his boss's discomfort at being probed as a political amphibian for supporting both Johnson for the nomination and liberal Democratic Senator Eugene McCarthy from Minnesota on the issues, and then having all of Kennedy's anguish about the war's tragic consequences pour out on national television.

Three days later, Johnson exiled Robert McNamara to the World Bank. Replacing him at the Pentagon was Clark Clifford, a leading hawk; having considered McNamara as the last voice in the cabinet for de-escalation and negotiation, Kennedy viewed Johnson as now unrestrained in his pursuit of the chimera of military victory.

That night, the newest and youngest addition to Kennedy's speech-writing team, twenty-four-year-old Jeff Greenfield, told his boss he had received his draft notice and didn't intend to serve. He expected Kennedy to tell him to leave the staff; instead, Kennedy joked, "Well, Jeff, you know, if you go to jail, I'll see to it that you get treated right—I used to have some influence [as attorney general] over the prison system . . . And besides, don't worry about it. A lot of the greatest men in history have begun their careers by spending time in jail."

On November 30, McCarthy, little known outside of the capital's Beltway, announced he would enter five or six primaries in 1968, as a focal point for antiwar sentiment. Kennedy's personal dilemma between politics and principle was no longer hypothetical.

Illusions in the Aftermath of Tet

CHICAGO, ILLINOIS

FEBRUARY 8, 1968

On January 23, the North Koreans seized the surveillance ship USS *Pueblo* and her eighty-three-man crew in international waters, twenty-five miles off the Korean coast.[18] In Kennedy's Senate office, Walinsky drew a practical political lesson from the incident: The president's inherent control of the foreign policy agenda stemmed in part from his ability to project American power at sites of his choosing; if he wished to provoke confrontations designed to rally domestic support, he had many opportunities to do so.[19] Nowhere were they more available than in Vietnam. Kennedy was well aware, in Walinsky's words, that "Johnson could step up the war, could provoke a wider war or could cut it all off." February began with a shock: As the Vietnamese Tet holiday began, the Vietcong launched a coordinated assault against American and South Vietnamese positions throughout the country. Saigon experienced fierce hand-to-hand fighting, and the American embassy was in enemy hands for several hours. In the first week of intensified action, nearly six hundred U.S. soldiers died. The Vietcong suffered heavier casualties and ultimately proved unable to hold the major cities they captured.

The administration declared it a decisive U.S. military triumph, but for many Americans, watching as the carnage unfolded on the nightly television news, the ability of the enemy to strike at will made it clear that there would be no quick victory over a foe that their military (and president) had repeatedly promised was near collapse.

Uncertain of what direction to take and angry at his indecision, Kennedy took Walinsky's insistence that the speech to the annual Book and Author luncheon on February 8 in Chicago (a key industry event originally selected to help promote *To Seek a Newer World)* address the war, advancing his existing proposals in light of the implications of the Tet Offensive. Walinsky decided to "let out all the stops," collaborating with Dick Goodwin (himself preparing to join McCarthy in New Hampshire, having decided he couldn't await a Kennedy candidacy any longer). Kennedy reviewed two alternatives for many of the key paragraphs; without exception, he elected those with the most direct criticism.

The events of the last few weeks have demonstrated anew the truth of Lord Halifax's dictum that although hope "is a very good company by the way . . .

[it] is generally a wrong guide." Our enemy, savagely striking at will across all of South Vietnam, has finally shattered the mask of official illusion with which we have concealed our true circumstances, even from ourselves. But a short time ago we were serene in our reports and predictions of progress. In April, our commanding general told us that "the South Vietnamese are fighting now better than ever before . . ." In August, another general told us that "the really big battles of the Vietnam war are over . . . the enemy has been so badly pummeled he'll never give us trouble again." In December, we were told that we were winning "battle after battle," that "the secure portion of the population has grown . . . and in contested areas the tide continues to run with us."

Those dreams are gone. The Vietcong will probably withdraw from the cities . . . thousands of them will be dead. But they will, nevertheless, have demonstrated that no part or person of South Vietnam is secure from their attacks: neither district capitals nor American bases, neither peasant in his rice paddy nor the commanding general of our own great forces . . .

Whatever their outcome, the events of the last two weeks have taught us something. For the sake of those young Americans who are fighting today, if for no other reason, the time has come to take a new look at the war in Vietnam; not by cursing the past but by using it to illuminate the future. And the first and necessary step is to face the facts. It is to seek out the austere and painful reality of Vietnam, freed from wishful thinking, false hopes, and sentimental dreams. It is to rid ourselves of the good company of those illusions which have lured us into the deepening swamp of Vietnam . . . It is time for the truth.

We must, first of all, rid ourselves of the illusion that the events of the past two weeks represent some sort of victory. That is not so . . .

Again it is claimed that the Communists expected a large-scale popular uprising which did not occur. How ironic it is that we should claim a victory because a people whom we have given sixteen thousand lives, billions of dollars, and almost a decade to defend, did not rise up in arms against us. More disillusioning and painful is the fact the population did not rise to defend its freedom against the Vietcong . . .

This has not happened because our men are not brave or effective, because they are. It is because we have misconceived the nature of the war; it is because we have sought to resolve by military might a conflict whose issue depends upon the will and conviction of the South Vietnamese people. It is like sending a lion to halt an epidemic of jungle rot.

This misconception rests on a second illusion—the illusion that we can win a war which the South Vietnamese cannot win for themselves. Two presidents and countless officials have told us for seven years that although we can help the South Vietnamese, it is their war and they must win it . . . Yet this wise and certain counsel has gradually become an empty slogan, as mounting frustration has led us to transform the war into an American military effort . . .

Every detached observer has testified to the enormous corruption which pervades every level of South Vietnamese official life . . . Despite continual promises, the Saigon regime refuses to act against corruption.

. . . Perhaps we could live with corruption and inefficiency by themselves. However, the consequence is not simply the loss of money or of popular confidence; it is the loss of American lives. For government corruption is the source of the enemy's strength. It is, more than anything else, the reason why the greatest power on earth cannot defeat a tiny and primitive foe.

You cannot expect people to risk their lives and endure hardship unless they have a stake in their own society. They must have a clear sense of identification with their own government, a belief they are participating in a cause worth fighting for. Political and economic reform are not simply idealistic slogans or noble goals to be postponed until the fighting is over. They are the principal weapons of battle. People will not fight to line the pockets of generals or swell the bank accounts of the wealthy. They are far more likely to close their eyes and shut their doors in the face of their government—even as they did last week . . .

We have an ally in name only. We support a government without supporters. Without the efforts of American arms, that government would not last a day.

The third illusion is that the unswerving pursuit of military victory, whatever its cost, is in the interest of either ourselves or the people of Vietnam. For the people of Vietnam, the last three years have meant little but horror . . . Nor does it serve the interest of America to fight this war as if moral standards could be subordinated to immediate necessities . . . Of course, the enemy is brutal and cruel, and has done the same thing many times. But we are not fighting the Communists in order to become more like them; we fight to preserve our differences . . .

The truth is that the war in Vietnam does not promise the end of all threats to Asia and ultimately to the United States; rather, if we only proceed on our present course, it promises only years and decades of further draining conflict on the mainland of Asia—conflict which, as our finest military leaders have always warned, could lead us only to national tragedy.

There is an American interest in South Vietnam. We have an interest in maintaining the strength of our commitments—and surely we have demonstrated that . . . And we have another, more immediate interest: to protect the lives of our gallant young men and to conserve American resources . . .

The fifth illusion is that this war can be settled in our own way and in our own time on our own terms. Such a settlement is the privilege of the triumphant; of those who crush their enemies in battle or wear away their will to fight.

We have not done this, nor is there any prospect we will achieve such a victory.

Once, in 1962, I participated in such predictions myself. But for twenty years we have been wrong. The history of conflict among nations does not record another such lengthy and consistent chronicle of error. It is time to discard so proven a fallacy . . .

Unable to defeat our enemy or break his will—at least without a huge, long, and ever more costly effort—we must actively seek a peaceful settlement. We can no longer harden our terms everywhere Hanoi indicates it may be prepared to negotiate; and we must be willing to foresee a settlement which will give the Vietcong a chance to participate in the political life of the country. Not because we want them to, but because that is the only way in which this struggle can be settled. No one knows if negotiations will bring a peaceful settlement, but we do know there will be no peaceful settlement without negotiations. Nor can we have these negotiations just on our own terms . . . What we must not do is confuse the prestige staked on a particular policy with the interest of the United States; nor should we be unwilling to take risks for peace when we are willing to risk so many lives in war . . .

These are some of the illusions which must be discarded if the events of last week are to prove, not simply a tragedy, but a lesson—a lesson which carries with it some basic truths . . .

The central battle in this war cannot be measured by body counts or bomb damage, but by the extent to which the people of South Vietnam act on a sense of common purpose and hope with those that govern them . . .

The best way to save our most precious stake in Vietnam—the lives of our soldiers—is to stop the enlargement of the war, and . . . the best way to end casualties is to end the war . . .

Our nation must be told the truth about this war, in all its terrible reality, both because it is right and because only in this way can any administration rally the public confidence and unity for the shadowed days which lie ahead.

No war has ever demanded more bravery from our people and our government—not just bravery under fire or the bravery to make sacrifices, but the bravery to discard the comfort of illusion, to do away with false hopes and alluring promises. Reality is grim and painful. But it is only a remote echo of the anguish toward which a policy founded on illusion is surely taking us. This is a great nation and a strong people. Any who seek to comfort rather than to speak plainly, reassure rather than instruct, promise satisfaction rather than reveal frustration—they deny that greatness and drain that strength. For today as it was in the beginning, it is the truth that makes us free.

Kennedy, remembered Walinsky (who accompanied him to Chicago), "*really* enjoyed giving [the speech]. You could see he really . . . got into it more and more, he really got very passionate about the delivery. And . . . when you're really biting and sharp about the delivery, it makes an even greater impact." The audience approved, but national reaction was mixed. Columnist Joseph Alsop told Kennedy of nonpartisan friends of Alsop's who said "they were compelled to regard Bobby Kennedy as a traitor to the United States." A Gallup poll taken in early February, while the meaning of the Tet fighting was not yet clear, showed nearly a three-to-one advantage for self-described hawks over doves, and a 70 percent rate of support for continued bombing.

Polls, however, were having a diminished influence on Kennedy. Jeff Greenfield noted that

> *there was obviously some change in the wind . . . The February 8 . . . speech . . . was wholly different in tone from anything he'd ever said on Vietnam—no attempt to preserve the bridges to the administration that was so much a part of the other speeches . . . The February 8 speech was simply an angry statement of a man who was simply disgusted with the way the policy was going.*

It was also the statement of a man increasingly convinced that the war would stop only when Lyndon

Johnson left the White House—and that he was the only Democrat who could hasten that departure.

Meanwhile, Lyndon Johnson and Eugene McCarthy neared their March 12 showdown in the snows of New Hampshire.

PART FOUR

The 1968 Presidential Campaign

Announcement of Candidacy for President

WASHINGTON, D.C.

MARCH 16, 1968

After a year of indecision, Robert Kennedy had decided to run for president. Political calculation had held him back: the conventional wisdom that one *never* challenged an incumbent president and fellow Democrat and never hatched a race one couldn't win. Equally worrisome had been concerns over press and public perception of the motivation for his candidacy. For years, journalists had treated each substantive difference with the administration as reflecting his ambition and his widely reported dislike of President Johnson, rather than being stimulated by a series of principled disagreements over the war and the neglected problems at home.

Two events in February 1968 were catalytic in Kennedy's decision-making: the Tet Offensive, which had demolished the administration's optimistic claims about the war, and Johnson's calculated indifference to the February 29 release of a provocative report from the Kerner Commission. Headed by Illinois governor Otto Kerner, the National Advisory Commission on Civil Disorders had been appointed by Johnson to analyze the 1967 urban riots. Its somber report described an

America divided, "moving toward two societies, one black, one white—separate and unequal." The White House ignored it. Having erred in Vietnam, the administration now had failed to respond to a challenge Kennedy deemed the country's central moral and practical domestic dilemma.

By March 3, he had decided to make the race and actively pursued support on a trip to the Midwest and California a week later. For a variety of personal and political reasons, he waited until after the New Hampshire primary, held on March 12, to announce. Eugene McCarthy polled 42.2 percent to 49.4 percent for the incumbent president (elected less than four years before by a landslide margin), and the senator from Minnesota was catapulted into the leadership of a growing national antiwar movement. In this context, Kennedy's announcement of his candidacy, a scant four days after McCarthy's "victory," seemed to justify the characterization he had strived to avoid: ruthless and opportunistic.

Kennedy announced his candidacy in the Senate Caucus Room, in the Capitol's old Senate Office Building. Here he had won his first national recognition during the televised hearings of the McClellan Committee on organized crime and labor racketeering. It was the same room in which his brother announced his presidential candidacy in January 1960 (then, John Kennedy had been forty-two, Robert Kennedy's age in 1968). Robert Kennedy, with his wife and children seated next to him, spoke to a packed room of reporters and staff.

I am announcing today my candidacy for the presidency of the United States.

I do not run for the presidency merely to oppose any man but to propose new policies. I run because I am convinced that this country is on a perilous course and because I have such strong feelings about what must be done, and I feel that I'm obliged to do all that I can.

I run to seek new policies—policies to end the bloodshed in Vietnam and in our cities, policies to close the gaps that now exist between black and white, between rich and poor, between young and old, in this country and around the rest of the world.

I run for the presidency because I want the Democratic Party and the United States of America to stand for hope instead of despair, for reconciliation of men instead of the growing risk of world war.

I run because it is now unmistakably clear that we can change these disastrous, divisive policies only by changing the men who are now making them. For the reality of recent events in Vietnam has been glossed over with illusions.

The report of the Riot Commission has been largely ignored.

The crisis in gold, the crisis in our cities, the crisis in our farms and in our ghettos have all been met with too little and too late.

No one who knows what I know about the extraordinary demands of the presidency can be certain that any mortal can adequately fill that position.

But my service in the National Security Council during the Cuban Missile Crisis, the Berlin crisis of 1961 and 1962, and later the negotiations on Laos and on the Nuclear Test Ban Treaty have taught me something about both the uses and limitations of military power, about the opportunities and the dangers which await our nation in many corners of the globe in which I have traveled.

As a member of the cabinet and a member of the Senate I have seen the inexcusable and ugly deprivation which causes children to starve in Mississippi; black citizens to riot in Watts; young Indians to commit suicide on their reservations because they've lacked all hope and they feel they have no future; and proud and able-bodied families to wait out their lives in empty idleness in eastern Kentucky.

I have traveled and I have listened to the young people of our nation and felt their anger about the war that they are sent to fight and about the world that they are about to inherit.

In private talks and in public, I have tried in vain to alter our course in Vietnam before it further saps our spirit and our manpower, further raises the risks of wider war, and further destroys the country and the people it was meant to save.

I cannot stand aside from the contest that will decide our nation's future and our children's future.

The remarkable New Hampshire campaign of Senator Eugene McCarthy has proven how deep are the present divisions within our party and within our country. Until that was publicly clear, my presence in the race would have been seen as a clash of personalities rather than issues.

But now that the fight is on and over policies which I have long been challenging, I must enter that race. The fight is just beginning, and I believe that I can win . . .

Finally, my decision reflects no personal animosity or disrespect toward President Johnson. He served President Kennedy with the utmost loyalty and was extremely kind to me and members of my family in the difficult months which followed the events of November of 1963.

I have often commended his efforts in health, in education, and in many other areas, and I have the deepest sympathy for the burden that he carries today.

But the issue is not personal. It is our profound differences over where we are heading and what we want to accomplish.

I do not lightly dismiss the dangers and the difficulties of challenging an incumbent president. But these are not ordinary times and this is not an ordinary election.

At stake is not simply the leadership of our party and even our country. It is our right to moral leadership of this planet.

In the portions of the speech not included in the above excerpts, Kennedy had maintained that his "candidacy would not be in opposition to [McCarthy], but in harmony," in hopes that doing so would reduce the appearance of ruthlessness and make the ultimate appeal to McCarthy's followers easier. The press corps, however, found the notion of a "harmonious" struggle between two proud men, each wanting to be president, highly impractical, and the press conference following the announcement was punctuated by questioning that made Kennedy uncomfortable. In answer to an early question, Kennedy said he was "not asking for a free ride"; it was clear from this launching of the campaign that the press would not be giving him one. He also knew that Lyndon Johnson would bend the powers of the incumbency and his titanic will to stop a Kennedy candidacy, and in short order it too became obvious that Eugene McCarthy did not regard himself as a mere steward for the return to Camelot.

Whatever uncertainties lay ahead politically, Kennedy had no hesitation concerning the substantive foundation of his nascent campaign. Adam Walinsky, who continued to serve as chief speech writer, reflected that

> *one point of genuine professional pride is that at no time during the campaign did we have to scrounge or scramble around for anything to say. The intellectual and programmatic focus of the platform on which [Kennedy] ran for the presidency had been thoroughly worked out in those years in the Senate. Not only worked out but it had been tested and gone through and had been consulted upon. And he had [it] inside him . . . it wasn't anything that was created and shoved on him and carpentered and said, "Here's your platform. And we've got this for these guys and this for those guys."*

Ending the War in Vietnam

KANSAS STATE UNIVERSITY

MARCH 18, 1968

The passion and commitment of the candidate to his vision were especially clear in Kennedy's opposition to the war in Vietnam, the centerpiece of his candidacy. His views frequently had been unpopular, and the real test of public acceptance in a campaign context came in an early swing through Kansas, centered on two long-scheduled campus stops. The trip followed appearances in St. Patrick's Day parades in New York and Boston, and on *Meet the Press,* where Kennedy faced continued questioning of his motives and strategy.

Kansas was (and still is) a largely rural state, not intrinsically hospitable to a liberal, New York critic of a war effort then in progress. "You know, these people are very conservative," Kennedy reminded Walinsky, directing him to elaborate (in the draft for the first Kansas campus speech) on the candidate's justifications for his antiwar position.

Well aware of the criticism of his motives, Kennedy's campaign was proceeding cautiously, and the traveling party was therefore shocked by the large crowd, waving homemade signs, that cheered Robert and Ethel Kennedy on their nighttime arrival in Kansas City to transfer planes, and that greeted his brief remarks with a frenzy of shrieks and applause—and then swamped Kennedy in a sea of Kansans struggling to shake his hand or merely to touch him. At Topeka's airport a short time later, the crowds were larger and the adulation even more intense. Kennedy, swamped by well-wishers, mounted an airline ramp, and through a bullhorn asked the crowd for help in the campaign ahead and in the even more difficult work of correcting the country's direction.

The next day was one of the most emotional of the entire campaign. The morning began at Kansas State University. When originally scheduled, months previously, the topic was to have been the problems of poverty; but after Kennedy announced his candidacy, he decided to speak on the war to 14,500 students and faculty jammed in the university fieldhouse, many perched in the rafters. McCarthy supporters and those of presumed Republican candidate Governor Nelson Rockefeller of New York were in evidence, but their critical banners were nearly lost in a sea of signs: WE LOVE YOU BOBBY, SOCK IT TO 'EM, BOBBY, and the like.

He began with jokes, both prepared and off the cuff, and the audience responded warmly. What followed, the maiden formal campaign address, was a review of the various elements of the turbulent and restless times, centering on an aggressive assault on Johnson's policy in Indochina. It was lengthy, reflecting Kennedy's instructions, because of the assumption that Kansas would be conservative and hostile.

I am . . . glad to come to the home state of a . . . Kansan[1] who wrote, "If our colleges and universities do not breed men who riot, who rebel, who attack life with all the youthful vision and vigor, then there is something wrong with our colleges. The more riots that come on college campuses, the better world for tomorrow." . . .

For this is a year of choice—a year when we choose not simply who will lead us but where we wish to be led; the country we want for ourselves, and the kind we want for our children. If in this year of choice we fashion new policies out of old illusions, we ensure for ourselves nothing but crisis for the future, and we bequeath to our children the bitter harvest of those crises.

For with all we have done, with all our immense power and richness, our problems seem to grow not less, but greater. We are in a time of unprecedented turbulence, of danger and questioning. It is at its root a question of the national soul. The president calls it restlessness, while cabinet officers and commentators tell us that America is deep in a malaise of the spirit—discouraging initiative, paralyzing will and action, dividing Americans from one another by their age, their views, and the color of their skins.

There are many causes. Some are in the failed promise of America itself . . . Another cause is in our inaction in the face of danger. We seem equally unable to control the violent disorder within our cities, or the pollution and destruction of the country, of the water and land that we use and our children inherit. And a third great cause of discontent is the course we are following in Vietnam, in a war which has divided Americans as they have not been divided since your state was called Bloody Kansas.

All this—questioning and uncertainty at home, divisive war abroad—has led us to a deep crisis of confidence: in our leadership, in each other, and in our very self as a nation.

Today I would speak to you of the third of those great crises: of the war in Vietnam. I come here, to this serious forum in the heart of the nation, to discuss this war with you; not on the basis of emotion, but fact; not, I hope, in clichés, but with a clear and discriminating sense of where the national interest lies.

I do not want—as I believe most Americans do not want—to sell out American interests, to simply withdraw, to raise the white flag of surrender. That would be unacceptable to us as a country and as a people. But I am concerned—as I believe most Americans are concerned—that the course we are following at the present time is deeply wrong . . .

I am concerned that, at the end of it all, there will only be more Americans killed; more of our treasure spilled out; and because of the bitterness and hatred on every side of this war, more hundreds of thousands of Vietnamese slaughtered; so that they may say, as Tacitus said of Rome: "They made a desert, and called it peace." . . .

Let us begin this discussion with a note both personal and public. I was involved in many of the early decisions of Vietnam, decisions which helped set us on our present path. It may be that effort was doomed from the start; that it was never really possible to bring all the people of South Vietnam under the rule of the successive governments we supported—governments, one after another, riddled with corruption, inefficiency, and greed; governments which did not and could not successfully capture and energize the national feeling of their people. If that is the case, as it well may be, then I am willing to bear my share of the responsibility, before history and before my fellow citizens.

But past error is no excuse for its own perpetuation. "Tragedy is a tool for the living to gain wisdom, not a guide by which to live."[2] Now as ever, we do ourselves best justice when we measure ourselves against ancient tests, and a good man yields when he knows his course is wrong, and repairs the evils. The only sin is pride . . .

If the government's troops will not or cannot carry the fight for their cities, we cannot ourselves destroy them. That kind of salvation is not an act we can presume to perform for them. For we must ask our government—we must ask ourselves: Where does such logic end?

If it becomes "necessary" to destroy all of South Vietnam in order to "save" it, will we do that too? And if we care so little about South Vietnam that we are willing to see the land destroyed and its people dead, then why are we there in the first place?

Can we ordain to ourselves the awful majesty of God—to decide what cities and villages are to be destroyed, who will live and who will die, and who will join refugees wandering in a desert of our own creation? If it is true that we have a commitment to the South Vietnamese people, we must ask: Are they being consulted? . . .

Let us have no misunderstanding. The Vietcong are a brutal enemy indeed. Time and time again, they have shown their willingness to sacrifice innocent civilians, to engage in torture and murder and despicable terror to achieve their ends . . .

We set out to prove our willingness to keep commitments everywhere in the world. What we are ensuring, instead, is that it is most unlikely that the American people would ever be willing to again engage in this kind of struggle. Meanwhile, our oldest and strongest allies pull back to their own shores, leaving us alone to police all of Asia . . .

Higher yet is the price we pay in our own innermost lives, and in the spirit of our country . . .

And whatever the costs to us, let us think of the young men we have sent there: not just the killed, but those who have to kill; not just the maimed, but also those who must look upon the results of what they do.

It may be asked: Is not such degradation the cost of all wars? Of course it is. That is why war is not an enterprise lightly to be undertaken, nor prolonged one moment past its absolute necessity. All this—the destruction of Vietnam, the cost to ourselves, the danger to the world—all this we would stand, willingly, if it seemed to serve some worthwhile end. But the costs of the war's present course far outweigh anything we can reasonably hope to gain by it, for ourselves or for the people of Vietnam. It must be ended, and it can be ended in a peace for brave men who have fought each other with a terrible fury, each believing that he alone was in the right. We have prayed to different gods, and the prayers of neither have been answered fully.

Now, while there is still time for some of them to be partly answered, now is the time to stop.

Kennedy's program for peace in Vietnam tracked proposals he had been advancing for more than two years: negotiation with the National Liberation Front and its participation in South Vietnam's political life; a de-escalation of the fighting; and reforms by the United States–supported government.

As the crescendo of roaring support from the crowd at KSU became deafening, *Look* photographer Stanley Tretick, who had covered Kennedy since the rackets hearings, momentarily abandoned his journalistic dispassion and screamed to Kennedy's delighted staff, "He's going all the way! He's going all the fuckin' way!"

Recapturing America's Moral Vision

UNIVERSITY OF KANSAS

MARCH 18, 1968

If Vietnam was the nation's central foreign dilemma, Kennedy saw the principal challenge at home to be improving the condition of the poor and disenfranchised, and mending the fraying moral fabric of a nation of historically unimaginable material wealth now dispirited and adrift. He forcefully introduced this theme on the second stop during his first real day of campaigning, speaking at the University of Kansas. At a school with sixteen thousand students, Kennedy drew a crowd of twenty thousand in the Phineas Allen Fieldhouse; as had been the case at Kansas State University, Kennedy's crowds were the largest in campus history. Journalist Jules Witcover described the scene: "Students stamped and clapped in rhythm as the candidate entered; an overflow crowd sitting on the shiny wood floor got up and swirled around him, leaving only a small island of officialdom waiting on folded chairs in the center. From high above, in the press section, it looked and sounded like some overly done scene from a Hollywood movie of a presidential campaign—the jumping girls, the screams ricocheting off the distant fieldhouse walls."

Kennedy again galvanized his audience with self-deprecating wit and an intense attack on the administration's policy in Vietnam, and as at KSU, drew energy from the crowd's reactions. In the *Los Angeles Times,* Washington bureau chief Robert Donovan wrote that "at times, his speeches at Kansas State and the University of Kansas seemed to have a spellbinding effect on the students."

Initially, Kennedy had intended to avoid any lengthy discussion of the war while at the University of Kansas, having focused on it in the day's earlier speech. But because the students at Kansas State had reacted with such surprising, visceral enthusiasm, Kennedy and Walinsky began (the aide remembered in a later oral history interview) "just taking pages wholesale out of the last speech and shoving them into the new one." Despite this, the KU address retained its focus on (and echoed the announcement speech's images about) those Americans who were being neglected in the political dialogue and bypassed by the nation's growing economy. Most of all, it kept its rhetorical core, exposing the hollowness of defining national progress by mere material accumulation.

I have seen these other Americans—I have seen children in Mississippi starving, their bodies so crippled by hunger; and their minds have been so destroyed for their whole life that they will have no future. I have seen children in Mississippi—here in the United States, with a gross national product of eight hundred billion dollars—I have seen children in the Delta area of Mississippi with distended stomachs, whose faces are covered with sores from starvation, and we haven't developed a policy so that we can get enough food so that they can live, so that their lives are not destroyed. I don't think that's acceptable in the United States of America and I think we need a change.

I have seen Indians living on their bare and meager reservations, with no jobs, with an unemployment rate of 80 percent, and with so little hope for the future that for young men and women in their teens, the greatest cause of death is suicide—that they end their lives by killing themselves.

I don't think that we have to accept that, for the first Americans, for the minority here in the United States. If young boys and girls are so filled with despair when they are going to high school and feel that their lives are so hopeless and that nobody's going to care for them, nobody's going to be involved with them, nobody's going to bother with them, that they either hang themselves, shoot themselves, or kill themselves—I don't think that's acceptable and I think the United States of America—I think the American people know we can do much, much better. And I run for the presidency because of that. I run for the presidency because I've seen proud men in the hills of Appalachia, who wish only to work in dignity, but they cannot, for the mines have closed and their jobs are gone and no one—neither industry, labor, nor government—has cared enough to help.

I think we here in this country, with the unselfish state that exists in the United States of America, I think we can do better here also.

I have seen the people of the black ghetto, listening to ever-greater promises of equality and justice, as they sit in the same decaying schools and huddle in the same filthy rooms, without heat, warding off the cold and warding off the rats.

If we believe that we, as Americans, are bound together by a common concern for each other, then an urgent national priority is upon us. We must begin to end the disgrace of this other America.

And this is one of the great tasks of leadership for us, as individuals and citizens this year. But even if we act to erase material poverty, there is another great task. It is to confront the poverty of satisfaction—a lack of purpose and dignity—that afflicts us all.

Too much and too long, we seem to have surrendered community excellence and community values in the mere accumulation of material things. Our gross national product now is over eight hundred billion dollars a year, but that GNP—if we should judge America by that—counts air pollution and cigarette advertising, and ambulances to clear our highways of carnage. It counts special locks for our doors and the jails for those who break them. It counts the destruction of our redwoods and the loss of our natural wonder in chaotic sprawl. It counts napalm and the cost of a nuclear warhead, and armored cars for police who fight riots in our streets. It counts Whitman's rifle and Speck's knife,[3] and the television programs which glorify violence in order to sell toys to our children.

Yet the gross national product does not allow for the health of our children, the quality of their education, or the joy of their play. It does not include the beauty of our poetry or the strength of our marriages; the intelligence of our public debate or the integrity of our public officials. It measures neither our wit nor our courage; neither our wisdom nor our learning; neither our compassion nor our devotion to our country; it measures everything, in short, except that which makes life worthwhile. And it can tell us everything about America except why we are proud that we are Americans.

If this is true here at home, so it is true elsewhere in the world. From the beginning, our proudest boast was that we, here in this country, would be the best hope for all of mankind. And now, as we look at the war in Vietnam, we wonder if we still hold a decent respect for the opinions of mankind, and whether they have maintained a decent respect for us; or whether, like Athens of old, we will forfeit sympathy and support, and ultimately security, in a single-minded pursuit of our own goals and our own objectives.

It was nearly impossible for Kennedy to return to his car through the throng of students; Jim Tolan, a campaign advance man, remembered: "It was the first time I was ever scared with him. Those kids were out of control. He could have got hurt, they liked him so much."

The Value of Dissent

VANDERBILT UNIVERSITY; NASHVILLE, TENNESSEE

MARCH 21, 1968

After the extraordinary campaign kickoff in Kansas, Kennedy barnstormed through Georgia, Alabama, and Tennessee in one day. Campuses, long a favored Kennedy venue, were not as natural a base of support in the South, where (according to polls) the Johnson administration's policy in Vietnam retained student confidence.

Campuses also reflected the tension between dissent and national unity that Kennedy strove to harmonize throughout his campaign. When his brother had been inaugurated on the frosty afternoon of January 20, 1961, thematically beginning the decade of the sixties, a confident and expectant nation thrilled to the young president's exhortations. But as Robert Kennedy sought the presidency in the year that marked the decade's true discordant climax, the public mood had become more grim and guarded, and the divided nation of the Kerner Commission's warning was splitting along more than merely racial lines.

These divisions were being exacerbated at both ends of the political spectrum: Some found even the most reasoned dissent treasonous, especially while the nation conducted (undeclared) war; while on the left (particularly on college campuses), dissent frequently degenerated into a denial of even the right to speak to those with differing views.

Kennedy felt passionately that dissent and debate must be encouraged—not choked off—in an atmosphere of tolerance and consideration if the country was to reach his stated goal of national reconciliation and rededication. In two of his speeches on March 21, he enunciated the sometimes difficult balance. At Vanderbilt University, the subject was the value of dissent.

When we are told to forego all dissent and division, we must ask: Who is it that is truly dividing the country? It is not those who call for change; it is those who make present policy who divide our country; those who bear the responsibility for our present course; those who have removed themselves from the American tradition, from the enduring and generous impulses that are the soul of the nation . . .

Those who now call for an end to dissent, moreover, seem not to understand what this country is all about. For debate and dissent are the very heart of the American process. We have followed the wisdom of Greece: "All things are to be examined and brought into question. There is no limit set to thought."

For debate is all we have to prevent past errors from leading us down the road to disaster. How else is error to be corrected, if not by the informed reason of dissent? Every dictatorship has ultimately strangled in the web of repression it wove for its people, making mistakes that could not be corrected because criticism was prohibited . . .

A second purpose of debate is to give voice and recognition to those without the power to be heard. There are millions of Americans living in hidden places, whose faces and names we never know. But I have seen children starving in Mississippi, idling their lives away in the ghetto; living without hope or future amid the despair on Indian reservations, with no jobs and little hope. I have seen proud men in the hills of Appalachia, who wish only to work in dignity—but the mines are closed, and the jobs are gone, and no one, neither industry or labor or government, has cared enough to help. Those conditions will change, those children will live, only if we dissent. So I dissent, and I know you do too.

A third reason for dissent is not because it is comforting but because it is not—because it sharply reminds us of our basic ideals and true purpose. Only broad and fundamental dissent will allow us to confront not only material poverty but the poverty of satisfaction that afflicts us all . . .

So if we are uneasy about our country today, perhaps it is because we are truer to our principles than we realize; because we know that our happiness will come not from goods we have but from the good we do together . . .

We say with Camus: "I should like to be able to love my country and still love justice." . . .

So I come here today, to this great university, to ask your help—not for me—for your country and for the future of your world. You are the people, as President Kennedy said, who have "the least ties to the present and the greatest ties to the future." I urge you to learn the harsh facts that lurk behind the mask of official illusion with which we have concealed our true circumstances, even from ourselves. Our country is in danger. Not just from foreign enemies; but, above all, from our own misguided policies, and what they can do to this country. There is a contest, not for the rule of America, but for the heart of America. In these next eight months, we are going to decide what this country will stand for, and what kind of men we are. So I ask for your help, in the cities and homes of this state, in the towns and farms, contributing your concern and action, warning of the danger of what we are doing, and the promise of what we can do. I ask you, as tens of thousands of young men and women are doing all over this land, to organize yourselves, and then to go forth and work for new policies—work to change our direction—and thus restore our place at the point of moral leadership, in our country, in our own hearts, and all around the world.

National Dialogue, National Reconciliation

UNIVERSITY OF ALABAMA

MARCH 21, 1968

His Vanderbilt message—that frank talk and free dissent were a virtue, particularly in those who would lead—became a frequent Kennedy campaign refrain. But the traits were a means to, not an end of, a politics of challenge and change aimed at creating a national consensus and then mobilizing citizen involvement in redirecting the country. As unsparing as Kennedy was in describing the terrible problems he saw within America and in its dealings abroad, he was idealistic and lyrical in urging his audiences to join him in reuniting their divided land. On the same day he spoke at Vanderbilt, Kennedy celebrated the shared hopes and necessary common action that could liberate the national spirit in a speech at the University of Alabama.

Almost five years before, Governor George Wallace had placed his body in the path of Kennedy's Justice Department deputy and a phalanx of federal marshals to try to stop the registration of two black students at the Tuscaloosa campus. Now, in the first week of his 1968 campaign, Kennedy appeared there before nine thousand students who had been expected to give no comfort to Wallace's nemesis. Kennedy's remarks instead were greeted warmly, a reaction that marked the entire southern campaign swing.

We speak as citizens of the same nation, joined by a common history, heartened by common success, troubled by common concerns. And I have come here to Alabama to talk with you of the hope which binds us together as Americans; and to ask for your help.

For America's successes were not built by men of narrow region, refusing to look beyond their own sectional concerns. The settling of the prairie by men of the East; the fight to build the Tennessee Valley Authority, led by George Norris of Nebraska and Franklin D. Roosevelt of New York; the battle of Alabama's Hugo Black for the rights of labor and free speech—these are triumphs for a whole nation, made by men who were first, and always, Americans. This is the spirit in which I came to Alabama.

I have come here because our great nation is troubled, divided as never before in our history; divided by a difficult, costly war abroad and by bitter, destructive crisis at home; divided by our age, by our beliefs, by the color of our skin. I have come here because I seek to join with you in building a better country and a united country. And I come to Alabama because I need your help.

For this campaign, in this critical election year, must be far more than a matter of political organization, of courting and counting votes.

This election will mean nothing if it leaves us, after it is all over, as divided as we were when it began. We have to begin to put our country together again. So I believe that any who seek high office this year must go before all Americans: not just those who agree with them, but also those who disagree; recognizing that it is not just our supporters, not just those who vote for us, but all Americans, who we must lead in the difficult years ahead.

And this is why I have come, at the outset of my campaign, not to New York or Chicago or Boston, but here to Alabama.

Some have said there are many issues on which we disagree. For my part, I do not believe these disagreements are as great as the principles which unite us. And I also think we can confront those issues with candor and truth, and confront each other as men. We need not paper over our differences on specific issues—if we can, as we must, remember always our common burden and our common hope as Americans . . .

For the work we must do is not for the benefit of

any one of our peoples: It is work we must do for all Americans . . .

All . . . Americans are joined by the bond of injustice—and all . . . Americans must be freed by strong, determined national effort—not an effort which merely swells our budget with programs which will not free [the less fortunate of] Americans—but an effort which will provide jobs, not welfare dollars; decent homes, not slums standing on the foundation of federal indifference . . .

For history has placed us all, northerner and southerner, black and white, within a common border and under a common law. All of us, from the wealthiest and most powerful of men to the weakest and hungriest of children, share one precious possession: the name American.

So I come to Alabama to ask you to help in the task of national reconciliation.

Johnson's Appeal to the Darker Impulses of the American Spirit

GREEK THEATRE; LOS ANGELES, CALIFORNIA

MARCH 24, 1968

On Saturday March 23, on the heels of the emotional Kansas trip and the subsequent southern barnstorming, Kennedy traveled to California, the scene of the season's climactic primary in June. Carl Greenberg of the *Los Angeles Times* wrote that "the reception Kennedy received [in Stockton, Sacramento, San Jose, Monterey, and Los Angeles] was uproarious, shrieking and frenzied."

Writing for the same paper, Robert Donovan described typical scenes:

> *Since the senator began campaigning . . . the soundwaves, if not the shockwaves, have been building up steadily . . . At Stockton, he was pulled off a car by crowds of admirers. At San Jose, an old woman tried to hug him and a young woman ruffled his hair . . . At Sacramento the crowd was so dense there was danger of panic . . . If the spectacle of the first nine days suggests anything, it is that the Kennedy candidacy has touched a live nerve. The people are troubled by*

war and racial tension. Especially in California, the reaction of the crowds was an indication that the Kennedy candidacy offered them some vague new hope. In San Jose, for example, an unmistakable gladness was in the air at his noisy reception.

By Sunday night, the pace of the schedule and the intensity of the public reaction brought the exhausted candidate to his last stop on this first swing through the Golden State. The evening marked only the second time on the California trip that Kennedy delivered a formal address from a prepared text, but it had not been drafted by his regular speech-writing team. Walinsky and Greenfield had been bumped off the chartered campaign plane to make room for contributors and politicians who wanted access to the candidate. Kennedy friend and frequent speech collaborator Dick Goodwin, still officially with the McCarthy campaign, provided Kennedy with a tough attack on Lyndon Johnson that the Minnesota senator had not used. Scanning the draft quickly on the boisterous plane, Kennedy felt comfortable with the rhetoric and substance; there was no time to reflect on nuance, and it was nuance that the press seized on that evening, for the campaign's first public miscue.

Six thousand people were jammed into the outdoor forty-five-hundred-seat Greek Theatre, and three thousand more listened outside, as Kennedy delivered his relatively brief remarks.

Surrounded as we are by crisis in Vietnam, civil strife in our great cities, and a division among our people, which often erupt in dramatic forms, it is easy to overlook the most profound crisis of all: the unprecedented and perilous drift of American society away from some of its most treasured principles.

This crisis is not dramatic. It does not suddenly flare into morning headlines or across the evening television screens. The movement cannot even be noticed as we go about our daily tasks. Yet over a period of years it has brought us to a most dangerous point. We know what this generation can accomplish.

We have had problems in the past. But at the same time we have shown that we can deal with our adversaries without bloodshed, as in the Cuban Missile Crisis. We know that we can move toward protecting mankind from nuclear disaster, as with the . . . Test Ban Treaty. We know that this nation can be fired by idealism and will serve the needs of others by peaceful means, as through the Peace Corps. We know we can begin to reduce the tensions between black and white, and not just through laws but personal leadership.

Together, we can make this a nation where young people do not seek the false peace of drugs. Together, we can make this a nation where old people are not shunted off; where, regardless of the color of his skin or the place of birth of his father, every citizen will have an equal chance at dignity and decency. Together, Americans are the most decent, generous, and compassionate people in the world.

Divided, they are collections of islands. Islands of blacks afraid of islands of whites. Islands of northerners bitterly opposed to islands of southerners. Islands of workers warring with islands of businessmen.

The sense of possibility matched to human capacity has been the central theme of our history, from the first settlers through Wilson and Roosevelt and Truman. It was the moving spirit of the Democratic Party in its proudest and most productive moments; it was the tradition of this country. Something has happened to that guiding spirit.

It is not just that policies are different or that I, and many others, disagree with what is being done. It runs far deeper than that. And much of that is that we have made a fundamental departure from the principles—

not only of the Democratic Party but of the country itself.

In specific terms the shift of deterioration is easier to see. It is most dramatically illuminated by the current disastrous course we follow in Vietnam. More than five hundred thousand American soldiers have been hurled into a bottomless Asian swamp against the counsel of almost every intelligent general from MacArthur to Ridgway. We know that by following the present course we cannot win a military victory—we cannot settle the war.

All that happens is ever-increasing destruction as frustration causes us to hurl more and more power against a small society . . . But the fact that our enemy is so primitive is also their greatest strength. For our power is meant to disable sophisticated, urban, technological societies. In Vietnam it is like fighting a swarm of bees with a sledgehammer . . .

All the phrases which have meant so much to Americans—peace and progress, justice and compassion, leadership and idealism—often sound not like stirring reminders of our nation, but call forth the cynical laughter or hostility of our young and many of our adults. Not because they do not believe them, but because they do not think our leaders mean them.

These specific failures reflect the larger failure of national purpose. We do not know where we are going. We have been stripped of goals and values and direction, as we move aimlessly and rather futilely from crisis to crisis and danger to danger. And the record shows that kind of approach will not only not solve problems, it will only deepen them.

This is not simply the result of bad policies and lack of skill. It flows from the fact that for almost the first time the national leadership is calling upon the darker impulses of the American spirit: not, perhaps, deliberately, but through its action and the example it sets—an example where integrity, truth, honor, and all the rest seem like words to fill out speeches rather than guiding beliefs. Thus we are turned inward. People wish to protect what they have. There is a failing of generosity and compassion. There is an unwillingness to sacrifice or take risks. All of this is contrary to the deepest and most dominant impulses of the American character—all that which has characterized two centuries of history.

The issue in this election, therefore, is whether this new and startling path shall continue into the future, or whether we shall turn back to our roots and to our tradition, so that future historians shall view this period as the great aberration of American history. That is the issue you must decide this year. That is why I am running. Not simply to become president of the United States. Not simply because I have new ideas and new programs and new policies. But because I hope to offer you in the form of my candidacy—because that is the only way our system allows such a choice—I hope to offer you a way in which the people themselves can lead the way back to those ideals which are the source of national strength and generosity and compassion of deed.

After half a century of increasingly more vicious personal attacks by candidates against rivals, it is difficult now to understand the uproar created by Kennedy's speech at the Greek Theatre, which never mentioned Johnson by name or title. In part, the criticism reflected the prevailing press corps belief that the president should not be the target of verbal attacks. To some extent, the stories were driven by writers opposed to Kennedy's views on the war. Other critics were exhibiting the political press's tendency to make a horse race out of an election by never letting any candidate get favorable coverage for too long. And, partially, some reporters were concerned that the candidate was unleashing uncontrollable emotionalism in the frenzied crowds that always surrounded him.

The *Los Angeles Times*'s Robert Donovan fired the campaign's first real press warning shot after the "Greek Theater" speech: "When a war becomes a flaming political issue, the line between debate and demagoguery becomes a thin one. A candidate can easily be carried across it in the ardor of the fight." Donovan reflected the views of some of the reporters traveling with Kennedy, whose concern had been building since the emotional reaction of the Kansas students. They felt that Kennedy had lost control of his own passion and had fueled the crowds' by moving from playfully engaging an audience (a Kennedy tradition) to manipulating it. Further, in Kansas the candidate provoked applause with a statement that some press critics labeled false: He had criticized the South Vietnamese for not drafting eighteen-year-olds when America demanded that sacrifice from her young. The Saigon government, in fact, had announced it would begin such a step, in large part provoked by Kennedy's criticism in his February 8 speech in Chicago. Now, after characterizing Johnson's perfidy as the root cause of the nation's many ills, Kennedy was accused of yielding to the temptation of demagoguery.

Kennedy and his staff believed that the campaign *had* liberated a spirit in the country, but that the released emotions reflected the empowerment of people no longer convinced that the president they had ceased trusting would inevitably win reelection, nor that the conflicts at home and abroad dividing and tormenting the nation were irreparable. "The sense of relief, of freedom, of breaking out," Walinsky later said about the attitude of the crowds in the early weeks of the campaign, "was just unbelievable. These audiences, they were frantic! I mean, three words and you could set them to cheering for minutes on end! And the way they grabbed at him and tore at him. He snapped the whole country out of it . . . they had so much more hope . . ."

In the final analysis, Kennedy (just as any public figure) could only capture an existing public mood or express a fundamental belief (albeit a repressed one). The notion of demagogic manipulation misses the essence of Kennedy's concept of public service; it also ignores, as Walinsky noted, that "*power* is a transitive verb. It requires an object . . . People have power to do certain things. And in that kind of situation a leader can have tremendous power to take people in certain directions and no power to take them in . . . a different direction that they don't themselves want to go."

After journalistic reaction to the Greek Theatre speech became manifest, the campaign and the candidate recalibrated: Fewer purely crowd events were scheduled; Kennedy's caution (which had tempered

A Candidate Confronts His Audience

IDAHO STATE UNIVERSITY

MARCH 26, 1968

INDIANA UNIVERSITY MEDICAL SCHOOL

APRIL 26, 1968

Kennedy's press critics, particularly in their coverage of the Greek Theatre speech, claimed he let his opposition to the Vietnam War and the adulation of the crowds lead him to incite his audiences deliberately. And yet no candidate in modern memory so often allowed the intensity of his moral outrage to provoke him into open confrontations with listeners whose support he had come to solicit. From the very beginning of his public career, Kennedy had made a point to venture into potentially hostile territory. During the 1968 campaign, he took this tendency further, with direct, impromptu encounters aimed at forcing his audiences out of their complacency. No longer would he merely engage them in a dialogue about broader societal issues; he now sought to force his listeners to examine their personal lives and attitudes, and he made little allowance for political calculation or bruised feelings.

Before his brother's assassination, Robert Kennedy had been a consummate political operative; as a candidate in the most important race of his career, political risks taken out of moral outrage shocked those who knew him only as a former campaign manager. Confrontations were relatively frequent, with an early example occurring only a week and a half into the campaign, when Kennedy spoke at Idaho State University, in Pocatello. To an audience of collegians relying on their student deferments to keep them out of the carnage in Vietnam, Kennedy bluntly and extemporaneously repeated his promise to stop a system that placed the burden of the fighting on the children of the poor and the working class.

We must realize that the system from which we have sent a disproportionate number of Negroes, Mexican Americans, and Indians to fight in Vietnam is a faulty one. That is why I support the abolition of the system as we know it. I as president would propose the establishment of a professional army.

With catcalls and clamor from the students interrupting him, Kennedy continued.

Now I know—I realize—I think I can understand your disappointment with my stand; however, while the war in South Vietnam is still taking place I am in favor of a lottery system. And student deferments should be abolished. The fact remains that the major burden of this war is being carried by the poor—the youths who can't afford college. As I mentioned, 10 percent of Americans are Negroes, yet 20 percent of Americans killed in Vietnam have been Negroes. In essence your chances of being killed there are twice as high if you are a Negro. What about the boy who simply wants to run a filling station, as opposed to one who wants to go to college? Why should he have to drop his plans and go? Is he not contributing to the betterment of our country with his determination and diligence? I say he is. We must change our system.

Even more confrontational was an appearance exactly a month later, before eight hundred medical students at Indiana University Medical Center's Emerson Hall Auditorium. The event's tone was evident immediately: During the introduction of the candidate, when a spectator in the balcony (later described variously as a black janitor or a student) shouted, "We want Bobby," a group of medical students in the front rows quickly responded, "No, we don't." Laughing with the audience, Kennedy noted: "They've been trying to form a doctors' committee for me in Indiana—and they're still trying." The students fired off questions challenging Kennedy on medical care for the poor, which they characterized as too expensive, unnecessary, or best left to the marketplace. Ultimately a student asked, "Where are you going to get all the money for these federally subsidized programs you're talking about?"

"From you," Kennedy fired back, and, advancing toward the students, he threw down the gauntlet:

Let me say something about the tenor of that question and some of the other questions. There are people in this country who suffer. I look around this room and I don't see many black faces who are going to be doctors. You can talk about where the money will come from . . . Part of civilized society is to let people go to medical school who come from ghettos. You don't see many people coming out of the ghettos or off the Indian reservations to medical school. You are the privileged ones here. It's easy to sit back and say it's the fault of the federal government, but it's our responsibility too. It's our society, not just our government, that spends twice as much on pets as on the poverty program. It's the poor who carry the major burden of the struggle in Vietnam. You sit here as white medical students, while black people carry the burden of the fighting in Vietnam.

He was interrupted by cries of: "We're going . . . we're going!"

Yes, but you're here now and they're over there. The war might be settled by the time you go.

Hisses and boos rang through the auditorium. Kennedy again noted how few black faces he saw in the audience, and a black student responded, "Hey, don't forget me." Kennedy replied, "I can see you, but you sure stand out." As the tension began to break, a student asked if the candidate would end medical school deferments; with a smile (but reflecting his often-stated position), Kennedy said, "The way things are going here today, probably yes." At the program's conclusion, a student stood up and stated that "a lot of us agree with what you are saying," and the candidate left to cheers and applause.

Jules Witcover, who had observed the colloquy, concluded that "it was the kind of give-and-take that [Kennedy] most enjoyed, and at which he was most effective. It also was an example of his inability to keep his emotional feelings in check all the time, even when the strategy called for him to woo the conservative vote."

Kennedy didn't doubt his stance's political cost. Turning to Walinsky as the two exited the hall, Kennedy quipped, "Well, we'll get lots of votes there!" At the very least, he knew that his frankness regarding taxes was, in Walinsky's words, certain to send "a chill up the spine of every orthopedist and orthodontist in the state of Indiana."

The Struggles of the "First Americans": RFK and Native Americans

ELEVENTH ANNUAL NAVAJO EDUCATION CONFERENCE; WINDOW ROCK, ARIZONA

MARCH 29, 1968

Kennedy's confrontation with the Idaho State students was part of an eight-state campaign sprint in the week following the Greek Theatre speech. (Of the states he visited, only Nebraska would be the site of a primary.) Although Kennedy's attacks on the president became less fierce in the face of the press criticism of the "darker impulses" address, the scheduling didn't, becoming even more grueling. (Some weeks later, when a reporter commented gratefully on a seemingly leisurely day with only a handful of stops, Kennedy joked, "I know; they fired Marat/Sade as head of scheduling.")

In the campaign's first fifteen days, he appeared in sixteen states (home to nearly two thirds of the convention delegates); he spoke to at least a quarter of a million people; and he received nearly nightly television news coverage, especially in the media markets he visited. Although an extraordinarily hardy man with a legendary reservoir of energy, Kennedy ended March requiring medical treatment for persistent laryngitis. Despite the rigors of his schedule (and the occasional misfiring by an organization so hastily convened), Kennedy was convinced that he understood the temper of his times and that his campaign was hitting its stride.

He was not shy about detours, however. In the midst of his frenetic late-March blitz, Kennedy adamantly rejected the urgings of his campaign advisers and took a small plane to tiny Window Rock, Arizona, to continue to show his solidarity with the American Indian.

Kennedy had been involved in Indian affairs as attorney general; the Justice Department's Lands Division (headed by Ramsey Clark, who became attorney general in 1966) represented the federal government in litigation over Indian land claims. Kennedy ended the traditionally reflexive federal opposition to these claims and personally intervened to facilitate a number of large claims settlements. He also spoke to Indian groups and others about the plight of the native peoples. In a 1963 address to the National Congress of American Indians, Kennedy said the Indian was "the victim of racial discrimination in his own land." Living conditions, particularly on reservations, were appalling: As Kennedy noted in a speech at Trinity College in Connecticut in 1963, the Indian "infant mortality rate . . . [was] nearly twice that of any other racial group in the country, and their overall life expectancy [was] twenty years less." Ramsey Clark noted that Kennedy "just had this burning passion to help people who had been denied justice."

In 1967, Kennedy persuaded the Senate Labor and Public Welfare Committee to create a Subcommittee on Indian Education. He became its chairman and traveled widely to reservations. Kennedy found few Indian teachers in the schools there, and virtually no materials on native culture and history. In subcommittee hearings and private discussions, including with officials of the Bureau of Indian Affairs, he tried to change conditions, especially regarding education and health care. And because he saw few politicians willing to involve themselves with Indians, he went out of his way to speak at tribal gatherings, in a show of solidarity. That spirit resulted in a departure from the 1968 campaign trail, to meet with the Navajo.

A senior staffer of the Indian Education Subcommittee had prepared a draft of an address for the occasion. Campaign speech writers Walinsky and Greenfield, accordingly, had anticipated taking the day off, but when they reviewed the suggested draft they immediately rejected it as too long: Greenfield remembered that "conservatively . . . [it] would have taken two hours to deliver . . . and would have just driven the senator into the ground." First sitting on an airport baggage cart with their typewriters on their laps, and then in narrow, bulkhead seats on the tiny plane flying them to Window Rock, the two went to work on a new speech, each taking half.

Kennedy, who had rested on the flight, had only a few moments to review the result on the way to the Window Rock Lodge Civic Center.

A crowd of four thousand awaited him there. Kennedy used his text only sparingly, speaking mostly extemporaneously.

Let me say this is a very distressing time—there are some very disturbing matters which I feel must be explored and remedied here in the Indian community.

The Indian is our First Citizen. The nurturing and emboldening of the young Indian's mind should be a priority to all of us here in America.

. . . But here, in Arizona, the United States government says we must send our youth to boarding schools—they are forced to go to them—often thousands of miles away from their parents. And this, at a time when a child suffers most from this separation. The role model of a parent—the ties that bind families across this nation—this is the strength, the fabric . . . [that] binds our country together, yet here this moral fiber is torn apart just as it is being kindled. I don't think that is acceptable . . .

What we're talking about is a national tragedy in the United States. Thousands of Indian children are out of school . . .

For this we cannot blame the Indian child, or the Indians' lack of motivation. I think it's about time we stopped blaming the Indian child for the large drop-out rate and put the blame where it belongs—on the system.

I believe this disgrace should last no longer. The Indian must have the power to shape his own education.

As recently as last winter two Navajo boys, feeling they could no longer bear the displacement of being herded to a boarding school, ran away, and rather than finding enlightenment in their freedom, they found death in the freezing air of the night.

This brutal tragedy is shocking evidence of America's failure to provide a decent educational program for Indian children. Furthermore, I believe this system—this shortsighted system—of Indian education is not really beneficial to the Indian youth. How could it be? This unresponsive system ignores the Indian culture and denies their heritage—their lineage—in essence, everything which makes Indians the proud, strong, gracious individuals they are, and what they mean to this country.

Too frequently this system is constructed around those doing the educating and not the Indian child. The purpose should be to help the child, not expand the bureaucracy . . . Navajo boarding schools crush the spirit of the child rather than embolden it. And this eventually defeats the purpose of education altogether.

This is a fight we're engaged in. A fight for accountability from the Bureau of Indian Affairs and the United States government. A fight which we must win—not for ourselves or our own satisfaction or justification—but for America's Indian children. This fight, however, will not be won by continuing to run unproductive boarding schools nor will the struggle be resolved by simply placing Indian children in public schools. We must dedicate ourselves to success in this effort, yet unless we allocate more funds directly to Indian education, we cannot hope for success.

But I think it should be clear that simply throwing money at any specific problem will not solve that problem. I believe we should supplement our ability to afford an increase in funds with an effort to recruit former Peace Corps and VISTA volunteers to aid in the instructional areas because they are uniquely qualified and would be able to gain a greater rapport with these young students . . . so these children will be prepared for the future and not misplaced by the past . . .

I believe it is time for the voice of the Navajo to be heard in the United States—not only heard but heeded.

Jeff Greenfield later recalled that Kennedy

> *was exhausted, and when he got really exhausted his voice took a somehow deeper quality. That is, it had a softer, very lyrical quality. When he began describing the treatment of the Indians, the front half of the hall was Bureau of Indian Affairs bureaucrats, and they sat on their hands. The back half of the hall was Indians, and they went—just so enthusiastic about the speech, I can't tell you . . . It was one of those acts of fate where it just happened to work out.*

Stopping Indecency at Home and Needless Sacrifices Abroad

PHOENIX, ARIZONA

MARCH 30, 1968

The campaign's opening weeks found major addresses, such as those at the two Kansas colleges, alternating with briefer stump speeches. A vivid example of the latter occurred the day after Kennedy spoke at Window Rock, when he addressed an outdoor rally at a shopping mall in Phoenix and voiced his two principal early campaign themes: that the nation had betrayed its essential decency by ignoring its less fortunate citizens, and that the flower of the country's youth was being sacrificed needlessly in Vietnam.

If there is one overriding reality in this country, it is that we must resist any erosion of a sense of national decency. Make no mistake: Decency is at the heart of the matter—and at the heart of this campaign. Poverty is indecent. Illiteracy is indecent.

The death or maiming of brave young men in the swamps of Asia, that is indecent.

It is indecent for a man to work with his back and

his hands without hope of ever seeing his son enter a university.

It is indecent for a man in the streets of New York or Portland, Detroit or Watts, to surrender the only life he will ever have to despair. It is indecent for the best of our young people to be driven to the terrors of drugs and violence, to allow their hearts to wither with hatred.

This is a time to create, not destroy. This is a time for men to work out of a sense of decency, not bitterness. This is a time to begin again.

And that is why I run for president, and that is why I ask for your help.

As we stand here today, brave young men are fighting across an ocean. Here, while the sun shines, men are dying on the other side of the earth.

Which of them might have written a great symphony? Which of them might have cured cancer? Which of them might have played in a World Series or given us the gift of laughter from a stage or helped build a bridge or a university? Which one of them would have taught a child to read?

It is our responsibility to let those men live. If they die because of our empty vanities, because of our failure of wisdom, because the world has changed and we have not changed with it, then we must answer to them.

And we must also answer to mankind.

We are the great country of diversity. There is room here for everyone. But our young people are our greatest natural resource . . .

We must bring them back into American life.

Statement on Johnson's Decision Not to Seek Reelection

OVERSEAS PRESS CLUB; NEW YORK, NEW YORK
APRIL 1, 1968

The Kennedy campaign's strategy initially was designed for a protracted struggle with Lyndon Johnson for the nomination. Party bosses, like Chicago Mayor Richard Daley, still controlled the nominating process. To demonstrate his electoral appeal to these power brokers, Kennedy sought to enter as many state primaries with diverse voter profiles as possible. The early scheduling (especially before the fallout from the Greek Theatre speech) gave ample opportunities for crowd events. Robert Donovan of the *Los Angeles Times* summarized the strategy in a March 25 article: "Kennedy's main objective is to stir up such tremendous excitement with an unprecedented springtime nationwide campaign that the shockwaves will jar loose Democratic delegates now aligned with the president . . . What Kennedy must do for the first time in modern American history is to build such a fire of popular sentiment under these delegates that they will be stampeded into supporting him instead of an incumbent president."

And although he avoided personal attacks on the incumbent, Kennedy often presented his vision of national problems and national potential by making Johnson his thematic and stylistic foil, hammering on the administration for the twin causes of national discord: Vietnam's bloody morass and America's yawning economic and racial divide.

And then, the opponent vanished.

Johnson concluded a televised address to the nation Sunday evening, March 31, with a sentence no pundit had predicted, and of which only a handful of family members and advisers had foreknowledge: "I shall not seek, and I will not accept, the nomination of my party for another term as your president."

New York Democratic Chairman John Burns rushed onto Kennedy's plane when it landed in New York and told him of Johnson's startling decision. Kennedy was surprised, and in the car riding into Manhattan he mused: "I wonder if he'd have done this if I hadn't come in." The campaign was in an uproar, and the scramble to phone key political leaders around the country began almost immediately and continued until nearly three o'clock in the morning.

Later that morning, before taking questions from reporters at New York's Overseas Press Club, Kennedy made the following statement:

Last night I sent the president the following telegram:

Mr. President:
First of all, let me say that I fervently hope that your new efforts for peace in Vietnam will succeed. Your decision regarding the presidency subordinates self to country and is truly magnanimous. I respectfully and earnestly request an opportunity to visit with you as soon as possible to discuss how we might work together in the interest of national unity during the coming months.

Sincerely,
Robert F. Kennedy

That wire sums up much of what I want to say today. The president's action reflects both courage and generosity of spirit. In these past sixteen days, I have been in some eighteen states, North, South, East, and West. In Alabama and in Watts, in New York and in New Mexico, in Washington, D.C., and in Washington State, wherever I went, I found Americans of all ages and colors and political beliefs deeply desirous of peace in Vietnam and reconciliation here at home. Despite all the discord and dispirit, despite all the extremists and their actions, there remains, in this country today, an enormous reservoir of hope and goodwill. Americans want to move forward; they want to better their communities, to make this country not only more livable for all Americans but a shining example for all the world.

To free their energies for progress at home, they want peace in Vietnam . . .

As we move toward a political resolution of the agony of Vietnam, we can start to redirect our national resources and energies toward the vital problems of our national community. The crisis of our cities, the tension among our races, the complexities of a society at once so rich and so deprived: All of these call urgently for our best efforts. We must reach across the false barriers that divide us from our brothers and countrymen, to seek and find peace abroad, reconciliation at home, and the participation in the life of our country that is the deepest desire of the American people and the truest expression of our national goals. In this spirit I will continue my campaign for the presidency.

Kennedy's proposal for a "national unity" meeting with the president was suggested by Ted Sorensen, one of President Kennedy's closest aides as White House special counsel and, in 1968, a New York lawyer with a lucrative practice. The campaign's "young turks" saw Sorensen as an influence for the status quo, and they felt any unity with Johnson would rob the campaign of its moral focus. The meeting between Kennedy (accompanied by Sorensen) and Johnson, and Johnson aides Walt Rostow and Charles Murphy, took place on April 3. Although cordial, it made clear what Kennedy suspected: Johnson had no intention of involving him in the peace process, of adopting Kennedy's views about the war, or of supporting his candidacy for president.

Child Poverty and Hunger

UNIVERSITY OF NOTRE DAME;
SOUTH BEND, INDIANA
APRIL 4, 1968

After Johnson's withdrawal, most of Kennedy's speeches centered on domestic issues. As a candidate attempting to appeal to the broad national electorate, however, he delivered few speeches about just one topic, instead tending to weave a variety of dangers and opportunities into a call for national renewal. He frequently spoke on agricultural policy (a critical concern in a number of the primary states), for example, and despite his frequent self-deprecating humor about his farm credentials—impeccable, he claimed, because of the tremendous consumption of dairy and other products by his enormous family—it was a subject on which he was well versed. He genuinely admired the dedication and community spirit of the small farmer.

This substantive range marked Kennedy's entire campaign, but where Vietnam had previously occupied center stage in the nation's political debate, after March 31 Kennedy focused on the plight of the impoverished. On April 4, 1968, at the University of Notre

Dame, before a capacity audience of five thousand, Kennedy used images that had haunted him from his Senate travels, and to which he frequently referred throughout the 1968 campaign, whether speaking off the cuff or from a prepared text.

This is the most affluent nation the world has ever known. This nation—our nation—has a food-producing capacity unrivaled in the history of the world. Yet, in the midst of our great affluence, children—American children—are hungry, some to the point where their minds and bodies are damaged beyond repair. I have seen, in the Mississippi Delta and on Indian reservations, children who eat only one meal a day—one meal of bread and gravy, or grits, or rice or beans. A distinguished Citizens' Board of Inquiry which has been studying the problem for nearly a year is reported to have found there are, conservatively, at least 10 million Americans suffering from hunger and malnutrition. This need not be the case. It must not be the case. If we cannot feed the children of our nation, there is very little we will be able to succeed in doing to live up to the principles which our founders set out nearly two hundred years ago.

The fundamental action we must take to feed the children of America is to see that their fathers have jobs—meaningful work so they can support their families with dignity and pride and buy the food that children need. If this is to be done in adequate measure, I believe the private enterprise system will have to play a major role, and I have introduced legislation which would make it possible for this to occur. I believe too that government should stand ready, on an immediate emergency basis, to employ or fund the employment of men who cannot find work. I have co-sponsored and long supported legislation to accomplish that end, and it should be enacted now . . .

Action in adequate measure can wait no longer. There are children in the United States of America with bloated bellies and sores of disease on their bodies. They have cuts and bruises that will not heal correctly in a timely fashion, and chronically runny noses. There are children in the United States who eat so little that they fall asleep in school and do not learn. We must act, and we must act now. And much of the action which must be taken needs no further authorization of Congress . . .

Beyond all this, it is time we designed a comprehensive program to feed our nation's hungry. The dominant purpose of the surplus commodities program is to dispose of surplus commodities so as to support farm prices. The dominant purpose of the food stamps program is to supplement what people normally spend for food. These purposes are sound in themselves, but there is no program whose avowed public purpose is to ensure that the hungry of our nation are fed and fed adequately. That is the purpose which must now be translated into concrete legislation . . .

These are our responsibilities. If we cannot meet them, we must ask ourselves what kind of a country we really are; we must ask ourselves what we really stand for. We must act—and we must act now.

During the same month, also in Indiana, David Murray of the *Chicago Sun-Times* witnessed Kennedy's special communion with the young, in a visit to a day nursery serving children of broken homes:

> *Two little girls came up and put their heads against his waist and he put his hands on their heads. And suddenly it was hard to watch, because he had become in that moment the father they did not know . . . You can build an image with a lot of sharpsters around you with their computers and their press releases. But lonely little children don't*

On the Death of the Reverend Dr. Martin Luther King Jr.

INDIANAPOLIS, INDIANA

APRIL 4, 1968

Kennedy launched his Indiana primary campaign on April 4, with a series of campus rallies. At Notre Dame, he had decried poverty, hunger, and joblessness. At Ball State University, he spoke of his belief in a generous America, and in the question-and-answer period that followed, a young black man asked, "You are placing great faith in white America. Is this faith justified?" Kennedy answered affirmatively and added, "I think the vast majority of white people want to do the decent thing." Immediately before boarding the plane in Muncie to fly to Indianapolis, Kennedy received a call from Pierre Salinger (his brother's press aide and a 1968 campaign adviser), who told him that Martin Luther King Jr. had been shot in Memphis.

Kennedy and his press secretary Frank Mankiewicz huddled on the short flight; they feared from the initial reports that King would not survive. Kennedy, still deeply scarred from his brother's assassination, struggled to comprehend this latest shooting and to determine what he should say about it publicly.

On landing in Indianapolis, Kennedy learned that King was dead. His advance men, Walter Sheridan and (now Congressman) John Lewis, had scheduled a rally in the heart of the city's ghetto, an event that Indianapolis Mayor Richard Lugar thought was too dangerous, even in the quieter times when the schedule was made.

The evening was chilly and raw. Lewis and Kennedy's traveling aides learned that the ghetto was quiet—word of the assassination had not reached the neighborhood—but Indianapolis Chief of Police Winston Churchill advised Kennedy not to go because he expected there would be violence as soon as the news became known. Kennedy grimly decided to proceed as planned, and, having sent his pregnant wife ahead to the hotel, he rode to the site in silent contemplation.

When his car entered the black neighborhood, the police escort dropped off. The press bus became separated from the candidate, and on the bus Frank Mankiewicz frantically tried to draft a coherent statement from the images and concerns Kennedy had expressed on the plane.

About one thousand people were waiting, in a carefree mood typical of a rally. Walinsky, also having been separated from Kennedy when the motorcade was disbanded, rushed up with suggested talking points. Kennedy thanked him distantly, crumpled the notes, and jammed them into his overcoat without glancing at them.

Having specifically requested that he not be formally introduced, Kennedy climbed onto a flatbed truck that was serving as a platform. "He was up there," said television correspondent Charles Quinn, "hunched in his black overcoat, his face gaunt and distressed and full of anguish." The only illumination came from floodlights, bathing the truck platform. As Mankiewicz raced up, with his own legal-pad speech draft, he realized he'd never reach the candidate in time.

Speaking extemporaneously, Kennedy said, in a sad, tremulous voice:

I have bad news for you, for all of our fellow citizens, and people who love peace all over the world, and that is that Martin Luther King was shot and killed tonight.

A huge gasp came from the people and then screams of "No! No!" Kennedy somberly went on.

Martin Luther King dedicated his life to love and to justice for his fellow human beings, and he died because of that effort.

In this difficult day, in this difficult time for the United States, it is perhaps well to ask what kind of a nation we are and what direction we want to move in. For those of you who are black—considering the evidence there evidently is that there were white people who were responsible—you can be filled with bitterness, with hatred, and a desire for revenge. We can move in that direction as a country, in great polarization—black people amongst black, white people amongst white, filled with hatred toward one another.

Or we can make an effort, as Martin Luther King did, to understand and to comprehend, and to replace that violence, that stain of bloodshed that has spread across our land, with an effort to understand with compassion and love.

For those of you who are black and are tempted to be filled with hatred and distrust at the injustice of such an act, against all white people, I can only say that I feel in my own heart the same kind of feeling. I had a member of my family killed, but he was killed by a white man. But we have to make an effort in the United States, we have to make an effort to understand, to go beyond these rather difficult times.

My favorite poet was Aeschylus. He wrote: "In our sleep, pain which cannot forget falls drop by drop upon the heart until, in our own despair, against our will, comes wisdom through the awful grace of God."

What we need in the United States is not division; what we need in the United States is not hatred; what we need in the United States is not violence or lawlessness; but love and wisdom, and compassion toward one another, and a feeling of justice toward those who still suffer within our country, whether they be white or they be black.

So I shall ask you tonight to return home, to say a prayer for the family of Martin Luther King, that's true, but more importantly to say a prayer for our own country, which all of us love—a prayer for understanding—and that compassion of which I spoke.

We can do well in this country. We will have difficult times; we've had difficult times in the past; we will have difficult times in the future. It is not the end of violence; it is not the end of lawlessness; it is not the end of disorder.

But the vast majority of white people and the vast

majority of black people in this country want to live together, want to improve the quality of our life, and want justice for all human beings who abide in our land.

Let us dedicate ourselves to what the Greeks wrote so many years ago: to tame the savageness of man and make gentle the life of this world.

Let us dedicate ourselves to that, and say a prayer for our country and for our people.

Kennedy was a unique white public official in America able to address a crowd in a black neighborhood that tragic night and not encounter violence. He spoke as a recognized champion of the disadvantaged, and he carried the credibility of his family's tragedy; in fact, this night marked the first time he had referred publicly to his brother's shooting.

Whether Kennedy's appearance was a factor—and it seemed surely to have been—Indianapolis remained quiet while rioting broke out in 110 cities across the land, causing thirty-nine deaths, twenty-six hundred injuries, and tens of millions of dollars of property damage.

Returning to his hotel, Kennedy phoned Coretta King, and then (at her request) asked Mankiewicz to make arrangements for the body of the slain civil rights leader to be flown from Memphis to King's home in

Atlanta. Carrier after carrier refused, anxious about the resulting notoriety, as the night of black rage continued to sweep across the nation. Finally, in the early morning hours, a private plane was borrowed from a Kennedy friend in Atlanta.

On the Mindless Menace of Violence

CLEVELAND, OHIO

APRIL 5, 1968

After King's death, Kennedy canceled all campaign appearances and withdrew to his Indianapolis hotel; however, various black leaders convinced him to keep a scheduled address to Cleveland's City Club the next day and to transform it into a plea for nonviolence. Both the killing of the Nobel Peace Prize winner and the rage and riots that followed his death were on Kennedy's mind as he asked his speech writers, Adam Walinsky and Jeff Greenfield (with the candidate on the road), to work by phone with his brother's principal draftsman, Ted Sorensen (in New York).

Kennedy himself was juxtaposing King's assassination with others from the turbulent decade, especially the death in Dallas. Normally after a day of campaigning, he would retire to his hotel room. The night of April 4, he sat up with his aides, although he was frequently withdrawn. He eventually left, then returned to the room where Walinsky and Greenfield were working. Sitting down on a bed, he was silent for a time, then said, "You know, that fellow Harvey Lee Oswald,[4] whatever his name is, set something loose in this country."

Greenfield and Walinsky, struggling to finish the draft, passed out over their typewriters in the predawn hours. Kennedy, uncharacteristically unable to sleep and (in Walinsky's words) "prowling around," happened upon his young aides, shut off the light, and sent them off for a few hours' rest. As he left Greenfield, the latter said to his boss, "You aren't so ruthless after all." With a smile, Kennedy replied, "Don't tell anybody."

The candidate reviewed the speech draft as he left for Cleveland, and editing changes were still under way as the plane landed. Despite its overnight, collaborative production, its anguished eloquence would be remembered two months later, when Kennedy himself lay dying.

To a hushed luncheon crowd, mostly white executives, Kennedy spoke softly but with obvious deep emotion.

This is a time of shame and sorrow. It is not a day for politics. I have saved this one opportunity, my only event of today, to speak briefly to you about the mindless menace of violence in America which again stains our land and every one of our lives.

It is not the concern of any one race. The victims of the violence are black and white, rich and poor, young and old, famous and unknown. They are, most important of all, human beings whom other human beings loved and needed. No one—no matter where he lives or what he does—can be certain who will suffer from some senseless act of bloodshed. And yet it goes on and on and on in this country of ours.

Why? What has violence ever accomplished? What has it ever created? No martyr's cause has ever been stilled by his assassin's bullet.

No wrongs have ever been righted by riots and civil disorders. A sniper is only a coward, not a hero; and an uncontrolled, uncontrollable mob is only the voice of madness, not the voice of reason.

Whenever any American's life is taken by another American unnecessarily—whether it is done in the name of the law or in the defiance of the law, by one man or a gang, in cold blood or in passion, in an attack of violence or in response to violence—whenever we tear at the fabric of the life which another man has painfully and clumsily woven for himself and his children, the whole nation is degraded.

"Among free men," said Abraham Lincoln, "there can be no successful appeal from the ballot to the bullet; and those who take such appeal are sure to lose their cause and pay the costs."

Yet we seemingly tolerate a rising level of violence that ignores our common humanity and our claims to civilization alike. We calmly accept newspaper reports of civilian slaughter in far-off lands. We glorify killing on movie and television screens and call it entertainment. We make it easy for men of all shades of sanity to acquire whatever weapons and ammunition they desire.

Too often we honor swagger and bluster and the wielders of force; too often we excuse those who are willing to build their own lives on the shattered dreams of others. Some Americans who preach nonviolence abroad fail to practice it here at home. Some who accuse others of inciting riots have by their own conduct invited them.

Some look for scapegoats, others look for conspiracies, but this much is clear: Violence breeds violence, repression brings retaliation, and only a cleansing of our whole society can remove this sickness from our soul.

For there is another kind of violence, slower but just as deadly destructive as the shot or the bomb in the night. This is the violence of institutions: indifference and inaction and slow decay. This is the violence that afflicts the poor, that poisons relations between men because their skin has different colors. This is the slow destruction of a child by hunger, and schools without books and homes without heat in the winter.

This is the breaking of a man's spirit by denying him the chance to stand as a father and as a man among other men. And this too afflicts us all.

I have not come here to propose a set of specific remedies, nor is there a single set. For a broad and adequate outline we know what must be done. When you teach a man to hate and fear his brother, when you teach that he is a lesser man because of his color or his beliefs or the policies he pursues, when you teach that those who differ from you threaten your freedom or your job or your family, then you also learn to confront others not as fellow citizens but as enemies, to be met not with cooperation but with conquest; to be subjugated and mastered.

We learn, at the last, to look at our brothers as aliens, men with whom we share a city but not a community; men bound to us in common dwelling but not in common effort. We learn to share only a common fear, only a common desire to retreat from each other, only a common impulse to meet disagreement with force. For all this, there are no final answers.

Yet we know what we must do. It is to achieve true

justice among our fellow citizens. The question is not what programs we should seek to enact. The question is whether we can find in our own midst and in our own hearts that leadership of humane purpose that will recognize the terrible truths of our existence.

We must admit the vanity of our false distinctions among men and learn to find our own advancement in the search for the advancement of others. We must admit in ourselves that our own children's future cannot be built on the misfortunes of others. We must recognize that this short life can neither be ennobled or enriched by hatred or revenge.

Our lives on this planet are too short and the work to be done too great to let this spirit flourish any longer in our land. Of course we cannot vanquish it with a program, nor with a resolution.

But we can perhaps remember, if only for a time, that those who live with us are our brothers, that they share with us the same short moment of life; that they seek, as do we, nothing but the chance to live out their lives in purpose and in happiness, winning what satisfaction and fulfillment they can.

Surely this bond of common faith, this bond of common goal, can begin to teach us something. Surely we can learn, at least, to look at those around us as fellow men, and surely we can begin to work a little harder to bind up the wounds among us and to become in our own hearts brothers and countrymen once again.

Kennedy left Cleveland for Washington, where fires raged throughout the nation's capital. Soldiers in full battle gear, deployed on tanks and armored personnel carriers, attempted to keep the curfew. On Sunday, April 7, he walked the riot-torn neighborhoods with D.C. City Councilman the Reverend Walter Fauntroy, who remembered that "the stench of burning wood and broken glass were all over the place. We walked the streets. The troops were on duty. A crowd gathered behind us, following Bobby Kennedy. The troops saw us coming at a distance, and they put on their gas masks and got the guns at ready, waiting for this horde of blacks coming up the street. When they saw it was Bobby Kennedy, they took off their masks and let us through. They looked awfully relieved."

The Differing Views of Racial Progress

TELEVISED PRESS CONFERENCE; WASHINGTON, D.C.

APRIL 7, 1968

Not since the heady days immediately following the Civil War had so much federal attention been focused on improving the conditions of black Americans as was placed by the Kennedy and early Johnson administrations. Important civil rights legislation (including the 1964 Civil Rights Act and the Voting Rights Act of 1965) had been passed, and the War on Poverty had begun. Nonetheless, riots tore through American cities in the summers of 1965 and 1967, and following King's death.

Many whites found black rage unappreciative at best and incomprehensible at worst. Many blacks found it baffling that whites failed to see that conditions for most American blacks, particularly those in urban ghettos, seemed worse in 1968 than they had been in 1960.

In response to a press conference question about the advisability of creating a domestic Peace Corps that would work in white suburban areas to try to reduce racism, Kennedy encapsulated the differing views from across the racial divide.

In the last five or six years, the white people have looked at the black people and said, "Look at all we have done. We passed the Civil Rights Act of 1964. We passed the [Voting] Rights Act of 1965.

"A Negro has been appointed to the Supreme Court. A Negro has been appointed to the cabinet. We passed the poverty program. We've done all of these things. And we spent all of this money; and yet there are riots in the cities. And there's lawlessness. And there's violence. And there's looting. Don't the black people understand what we've tried to do and that we've made this commitment and can't they be satisfied?"

The black person, on the other hand, says, "That's all fine for Mr. Weaver who's in the cabinet. And that's fine for Mr. Marshall who's on the Supreme Court. And that's fine for the civil rights bills that have been passed. But none of that has any effect on my life. The fact is my children still go to substandard schools. The fact is my husband can't get a job. The fact is that I can't get welfare unless I divorce my husband and my children are illegitimate."

The fact is that the housing is substandard and becoming more substandard. And the fact is, as the Kerner Commission and the Riot Commission reported, the conditions in the ghetto and the conditions for the poor are getting worse, not better.

So that just to have the black people understand the fact that white people want to do what is right and feel that they have taken steps may not be sufficient. But basically there is this [need] of generosity and compassion to have the white people understand that the conditions are still very difficult for black people.

The next day, Robert and Ethel Kennedy went to Atlanta, at Coretta King's request, for her husband's funeral. Hosea Williams, one of King's aides, summed up the feelings of much of the nation's black community: "We felt as long as Dr. King lived, he would lead us to higher grounds . . . But after he was killed, it left us hopeless, very desperate, dangerous men. I was so despondent and frustrated at Dr. King's death, I had to seriously ask myself . . . Can this country be saved? I guess the thing that kept us going was that maybe Bobby Kennedy would come up with some answers for this country . . . After Dr. King was killed, there was just about nobody else left but Bobby Kennedy. I remember telling him he had a chance to be a prophet. But prophets get shot."

Achieving Racial Understanding

FORT WAYNE, INDIANA

APRIL 10, 1968

The days following King's assassination posed maximum danger from racial tensions but also the greatest opportunity for inspiring sustained white attention to achieve reconciliation. In the third and final speech directly inspired by King's death, Kennedy moved beyond the calls for compassion and an end to violence, to an exhortation to his fellow whites finally and fully to share American society with their black fellow citizens.

It was a subject much on Kennedy's mind in the wake of King's death, as his April 7 press conference indicated. And the dire need to reach the average white worker was the topic of a meeting he had with aides late the following night, after the civil rights leader's funeral. Kennedy had long believed that whites must be made stakeholders in the fight for equal justice and had focused on the topic in the third of his series of speeches on the urban crisis in January 1966 and frequently thereafter. In Kennedy's mind, after King's death, the need for a strengthened commitment to racial understanding became nearly desperate. As Walinsky remembered, the central question became "How do you begin to engage the great middle class in a kind of public endeavor that among other things allows you to do something serious about the conditions of the black poor? . . . It had to be a spiritual, or personal . . . kind of involvement and commitment to the public purposes of the country."

Returning to Indiana, a stronghold for George Wallace four years earlier, Kennedy did not hesitate to make the challenge to the nation's (and that state's) majority crystal clear.

This generation did not create most of the conditions and convictions which have led us to this day, but this generation has the responsibility to resolve them. Leaders can explain and propose, but this problem will not yield to any man, even the president of the United States. It will yield only to the moral energy and belief of a free people.

Most people, of both races, do not wish violence and agree that we must do everything we can to protect the life and property of our people. Of course this is so. Yet while the necessary troops and police patrol our streets, we must also be aware that punishment is not prevention, nor is an armed camp a place of peace. Our nation today is beset by apprehension and fear, anger and even hatred. It is easy to understand the springs of such passion; even as we know the highest traditions of this country forbid them. But today's difficult issue is not whether white Americans will help black Americans, but whether we will help ensure the well-being of every citizen. It is not whether white and black will love one another, but whether they will love America. It is not whether we will enforce the laws of the nation, but whether there is to be one nation . . .

There is no sure way to suppress men filled with anger who feel they have nothing to lose . . .

It is too easy to say that other minorities found their own way, even if that were completely true. For we must remember that the progress of many groups, including my own ancestors, the Irish, was not unmarked by violence, repression, and hatred. But none of these groups had a heritage which helped destroy tradition and culture and in which it was dangerous for a black man to presume to learning. Gradually and painfully Negroes themselves are overcoming this legacy, but it is not easy. Other minorities got their start by going west or working as laborers, so their sons could study to become mechanics, and their sons could become lawyers or even politicians. Today we live in a vast and complicated world of sprawling cities and huge industries, all of it moved by intricate technology and elaborate organization. It takes help for a poor man in a black ghetto to find a place in such a world. In fact, it is hard for all the rest of us. And, too, this is the only American minority with a black skin. I doubt there are many of us who, if they look honestly into their own hearts, will not admit that this makes a difference.

So there is our problem. Among us are millions who wish to be part of this society—to share its abundance, its opportunity, and its purposes. We can deny this wish or work to make it come true. If we choose denial then we choose spreading conflict, which will surely erode the well-being and liberty of every citizen and, in a profound way, diminish the idea of America. If we choose fulfillment it will take work, but we will choose to improve the well-being of all our people; choose to end fear and heal wounds; and we will choose peace, the only peace that can last—peace with justice.

Once we make this choice, we know how to begin . . . [remembering that] laws and programs are only part of any answer. It is necessary to find ways to reopen and deepen the channels of communication between white and black America; and to halt the dangerous drift toward isolated enmity which may soon find us looking at each other across impassable barriers of suspicion and anger. This is a job for every American. At every level of national life from the White House to the local church and meetinghouse and in the homes of individual citizens, men and women of both races can try to meet and discuss their problems as well as their fears.

They would meet not merely in search of understanding—although that is important—but to find ways to devote their energy to this most critical American problem.

And in a deeper way, only through such shared enterprise can we demonstrate our willingness to open doors to the larger society—American society—which now seems so forbidding.

Income and education and homes do not make a nation. Nor do land and borders. Shared ideals and principles, joined purposes and hopes—these make a nation. And that is our great task: to make one nation out of two.

It is not written in the stars or ordained by history that we must be a nation torn by strife and teeming with troops and bayonets. It is not inevitable that each day—as I experienced last weekend in Washington—we must hear the constant shriek of sirens or see smoke rising from burning buildings.

Americans have always believed that if we faced our problems and worked at them, they could be resolved. And for the most part, they have been right. Those who now believe we have the power to do justice and to make our streets fit places where men can live and children play in tranquillity, are also right.

The enemies of such an achievement are not the black man or the white man. The enemies are fear and indifference. They are hatred and, above all, letting momentary passion blind us to a clear and reasoned understanding of the realities of our land.

Here we are, white and black, together on one continent. If we can live together, the land will be fruitful for all its people. If not, much that we all cherish may be in danger. These are the only practical and realistic choices.

The Condition of Blacks in America

MICHIGAN STATE UNIVERSITY

APRIL 11, 1968

Kennedy continued his effort to cajole, inspire—or shame—the commitment of the white majority. Whereas "his speeches at black rallies were short," campaign historian Theodore White noted, "the fury and indignation [Kennedy] felt at the condition of blacks in America he spent rather at university campuses or excoriating white audiences for their indifference."

The day after his speech in Fort Wayne, Indiana, Kennedy took his message beyond that first primary battleground state, and except for a brief stop on April 15, he did not return to the state for nearly two weeks. It was during this period of hopscotch around the country that he met with a predominantly white group of student leaders at an airport reception near Michigan State University, one week after Dr. King's death.

At the outset we must make it unmistakably clear: A violent few cannot be permitted to threaten the lives and well-being of the many, nor lawless gangs disrupt the peace of our cities and the hopes of their fellows for progress. But history offers cold comfort to those who think grievance and despair can be subdued by force.

To understand is not to permit. But to fail to understand is the surest guarantee of a mounting strife which will assault the well-being of every citizen. And therefore . . . the first task before us, is this: an effort to understand. For the division between black and white is not the result of a failure of compassion, or of the American sense of justice. It is a failure of communication and vision.

We live in different worlds and gaze out over a different landscape. Through the eyes of the white majority, the Negro world is one of steady and continuous progress. In a few years, he has seen the entire structure of discriminatory legislation torn down. He has heard presidents become spokesmen for racial justice, while black Americans enter the cabinet and the Supreme Court. The white American has paid taxes for poverty and education programs, and watched his children risk their lives to register voters in Alabama. Seeing this, he asks, What cause can there be for violent insurrection, of dissatisfaction with present progress? But if we try to look through the eyes of the young slum dweller—the Negro, and the Puerto Rican, and the Mexican American—the world is a dark and hopeless place indeed.

Let us look for a moment. The chances are that he was born into a family without a father, often as a result of welfare laws which require a broken home as a condition for help. I have seen, in my own state of New York, these children crowded with adults into one or two rooms, without adequate plumbing or heat, each night trying to defend against marauding rats. The growing child goes to a school which teaches little that helps him in an alien world. The chances are seven of ten that he will not graduate from high school; and even when he does, he has a fifty-fifty chance of acquiring only as much as the equivalent of an eighth-grade education. A young college graduate who taught in a ghetto school sums it up this way: "The books are junk, the paint peels, the cellar stinks, the teachers call you nigger, the windows fall in on your head."

Most important, the people of the ghetto live today with an unemployment rate far worse than the rest of the nation knew during the depth of the Great Depression . . .

And let us be clear that all this is true despite the laws, despite the programs, despite all the speeches and promises of the last seven years. It must be for us a cruel and humbling fact—but it is a fact nonetheless—that our efforts have not even maintained the problem as it was: Economic and social conditions in these areas, says the Department of Labor, are growing worse, not better.

But this is not all the young man of the ghetto can see. Every day, as the years pass, and he becomes aware that there is nothing at the end of the road, he watches the rest of us go from peak to new peak of comfort. A few blocks away or on his television set, the young Negro of the slums sees the multiplying marvels of white America: more new cars and more summer vacations, more air-conditioned homes and neatly kept lawns. But he cannot buy them.

He is told that Negroes are making progress. But what can that mean to him? He cannot experience the progress of others, nor should we seriously expect him to feel grateful because he is no longer a slave, or because he can vote, or eat at some lunch counters. He sees only the misery of his present and the darkening years ahead. Others tell him to work his way up as other minorities have done; and so he must. For he knows and we know, that only by his own efforts and his own labor will the Negro come to full equality.

But how is he to work? The jobs have fled to the suburbs or been replaced by machines, or have flown beyond the reach of those with limited education and skills . . .

And thus, the black American youth is powerless to change his place or to make a better one for his children. He is denied the most fundamental of human needs: the need for identity, for recognition as a citizen and as a man.

Here, and not in the pitiful charade of revolutionary oratory, is the breeding ground of reverse racism, and of aimless hostility and of violence. The violent youth of the ghetto is not simply protesting his condition but making a destructive and self-defeating attempt to assert his worth and dignity as a human being, to tell us that though we may scorn his contribution, we must still respect his power . . .

We must now commit ourselves to the proposition that as funds begin to be released from Vietnam, they will come home to the service of our domestic peace. We have to begin planning now the rebuilding effort to come. And if, from the outset, we seek the full participation of the people of the other America in this initial effort, then I believe they will understand, as we know, that everything cannot be done at once. They want not promises which cannot be met, but genuine commitment to achievable goals. We must make that commitment, and plan seriously for its fulfillment with them. That is a place to begin.

Let us, then, turn to our cities, where so much is promised, so much undone, so much threatened. There is work here for all—in the service of the most urgent public needs of the nation; work that will benefit white and black, fortunate and deprived alike. City hospitals and school classrooms are overcrowded and outdated everywhere; tens of thousands of young men and women cannot attend college because there simply is no room. In fact, the inventory is almost infinite: parks and playgrounds to be built, public facilities to be renovated, new transportation networks to be established, rivers and beaches to be cleansed of filth and again made fit for human use. And to all this must be added the demand of an expanding economy for a growing nation.

Now, with all this to be done, let us stop thinking of the poor—the dropouts, the unemployed, those on welfare, and those who work for poverty wages—as liabilities. Let us see them for what they are: valuable resources, as people whose work can be directed to all these tasks to be done within our cities and within the nation.

Community, Compassion, and Involvement

SCOTTSBLUFF, NEBRASKA

APRIL 20, 1968

Whether or not his white student audience in Michigan rose to compassion and comprehension, Kennedy nonetheless exhibited a consistent appeal to working-class, ethnic whites, who perceived themselves as most threatened by blacks and who were the most hostile in their racial attitudes. In 1964, George Wallace had run an openly racist campaign in Indiana and garnered nearly a third of the vote. Television correspondent Charles Quinn related having a number of conversations around the country with white ethnics (some, fervent Wallace supporters) such as this one, in Gary, Indiana:

Q.—Well, what do you think of Kennedy?
A.—We like Kennedy very much . . .

Q.—But I understand you're not terribly crazy about Negroes.
A.—Naw, don't like Negroes. Nobody around here likes Negroes.

Q.—Here's a man who stands for helping the Negro, and you say you don't like them. How can you vote for him?

A.—I don't know. Just like him . . .

Quinn concluded that

all these whites, all these blue-collar people and ethnic people who supported Kennedy . . . as they did in Indiana and in Nebraska . . . and as they did to a lesser degree in California . . . all of these people felt that Kennedy would really do what he thought was right for the black people but, at the same time, would not tolerate lawlessness and violence. The Kennedy toughness came through on that. They were willing to gamble . . . because they knew in their hearts that the country was not right . . . they were willing to gamble on this man, maybe, who would try to keep things within reasonable order; and at the same time, do some of the things that they knew really should be done.

In part for the same reasons, Kennedy found a receptive audience among whites in rural areas as well. The feeling was reciprocated: Kennedy told Peter Edelman that he felt most comfortable in small-town America.

Echoing others, Edelman believed that to a certain extent the same romantic and unexpected identification that arose between John Kennedy and West Virginia coal miners developed between Robert and Nebraska farmers. Before visiting the state during the campaign, "Robert Kennedy hadn't known what terrible problems the farmers had, and then he went there, and really saw it in their faces and on the land . . . he saw . . . the farmer being another kind of forgotten and alienated American, another person who thought that this system had just left him behind."

That shared characteristic allowed Kennedy to forge a coalition of urban blacks, ethnic working-class whites, and farmers; and he fervently believed in uniting them in a recommitment to shared principles of compassion, community, and empowerment. He outlined his vision in mid-April, at a high school in the small town of Scottsbluff, Nebraska, in the midst of a barnstorming day of campaigning in that state.

John Adams once said that he considered the founding of America part of "a divine plan for the liberation of the slavish part of mankind all over the globe."

This faith did not spring from grandiose schemes of empires abroad. It grew instead from confidence that the example set by our nation—the example of individual liberty fused with common effort—would spark the spirit of liberty around the planet; and that once unleashed, no despot could suppress it, no prison could restrain it, no army could withstand it.

It is easy to forget, in the midst of a difficult struggle abroad that we have not won, and a struggle at home that we have not begun—it is easy to forget that the history of America is in large measure the redemption of the faith of the Founding Fathers.

For at every critical mark in our history, Americans have looked beyond the narrow borders of personal concern, remembering the bonds that tied them to their fellow citizens . . .

These efforts were not acts of charity.

They sprang from the recognition of a root fact of American life: that we all share in each other's fortunes; that where one of us prospers, all of us prosper; and where one of us falters, so do we all.

It is this sense, more than any failure of goodwill or of policy, that we have missed in America.

As our nation—and its problems—have grown, we seem to have grown apart from one another.

We seem, through no fault of our own, to look only the short distance; to turn away from the far horizon; to work, each of us, on building a piece of our country.

And the pieces do not match . . .

We became separated from one another, treating those of different races, or religions, or calling, as adversaries instead of allies . . .

The first step in this task is to remember what our government should be . . . a reflection of a common effort, a means of aspiring greater individual opportunity for our citizens . . .

The goal of an active federal effort in our social dilemmas must be to let loose the talent and energy of the people themselves, not to channel that energy along rigidly preconceived paths.

What we seek is not just greater programs but greater participation—by putting our resources directly into communities, both urban and rural, where the citizenry can determine how best to use those resources.

It is true that town-hall America may be gone; but that is no reason why small groups of people cannot plan for their own future, and decide their own fate, if government remembers how effective citizen participation can be.

It is time—indeed, the time is long since past—that the government begin to accommodate itself to the requirements of its citizens, instead of the other way around.

There is nothing sacred about government procedure.

There is nothing irrevocable about an old structure which forces small communities without the resources to hire lobbyists, to deal with twenty different federal agencies in a frustrating, fruitless search for assistance.

Here is where executive leadership can help—and where it will help if I am the next president.

The final necessity is for understanding. No leadership can separate itself from the people's hopes and wants. For the first task of any new leadership will be to rally the diverse forces within America to that common effort all of us require.

That is why, if I am chosen as your president, I pledge to you to go out among you: to meet with you, not as a speaker, but as a listener; to open again those channels of communication so vital to a democracy.

So, if I become your president, I intend to travel regularly across America, talking with the people of our country . . . I intend not only to make government more responsive to the people, but to reach out beyond the apparatus of government, talking directly with the people themselves and giving them the opportunity to talk with me.

That is the kind of leadership we need. That is the

leadership I intend to offer. There is much to be done in America. But with leadership confident in our citizens, these tasks can be done, this country can be put back together, and we shall in fact become "the last, best hope of man."

The Conduct of American Foreign Policy

UNIVERSITY OF INDIANA; BLOOMINGTON, INDIANA

APRIL 24, 1968

Kennedy asked his audience in Scottsbluff, and countless others across the country, to join him in direct citizen involvement in national renewal and, implicitly, in an unswerving popular assault on the Democratic Party apparatus controlling the nominating process. Another prong of Kennedy's challenge to the party establishment continued to be his opposition to the war in Vietnam, where citizen involvement had helped to force the incumbent out of the race. And although the debate over that conflict was dampened by Johnson's March 31 decision to not seek reelection (and the simultaneous announcement of the administration's intention to begin the peace process), Kennedy continued to speak against the war throughout the campaign.

One address, delivered nearly a month after Johnson's announcement (and only three days before Vice President Humphrey declared his candidacy), was particularly notable for applying the country's experience in Vietnam to reorient broader principles of American foreign policy around what Kennedy identified as the nation's actual and legitimate interests. The speech also utilized Kennedy's nationwide network of experts:

Drafted by Walinsky based on Kennedy's directions, it was transmitted to Edelman in Washington, who vetted it with a host of foreign policy specialists, in and out of government. Kennedy chose the University of Indiana, in Bloomington, as his venue, two days into an intensive sixteen-day swing through the state that would climax on primary day, May 7.

Long ago it was said, "The time for taking a lesson from history is ever at hand for those that are wise." The war in Vietnam is not yet consigned to history. The fighting and bloodshed continue. The bombing of North Vietnam is restricted; but that too continues. And the negotiations, toward which we have taken the most tentative and still far from certain steps—these have not yet begun.

Still, in one sense the war may be passing into history; and that is in the thinking of the American people. There has been settled, in the year 1968, one simple proposition: The American people—scholars and officials, soldiers and citizens, students and parents—are determined that there must not be another Vietnam . . .

What does it mean to say, "No more Vietnams"?

It is sometimes asserted that Vietnam is the battlefield where we must make the decision which may well determine the future shape of Asia. But it is unlikely that whatever the outcome of the war in Vietnam, the dominoes will fall in either direction . . .

Kennedy then reviewed the various regions of the world, concluding that Communists were unlikely to be able to impose a government anywhere they did not already control.

All this is not to say that Vietnam is the last challenge we will face abroad. That would be nonsense. It is to say that Vietnam is only Vietnam—that it will not settle the fate of Asia or of America, much less the fate of the world.

But it is also true that the danger is not past. Almost all the nations of Asia and Africa are only recently emerged from colonial domination, from three hundred years in which the entire structure of their societies and culture was torn apart, degraded, and humiliated. They are still torn today by the tremendous effort to modernize and develop their economies; to create new leadership groups capable of managing a modern society and to cope with demands for social justice that have been awakened by the example of the successful egalitarian West. We can expect that these nations and the nations of Latin America, which are Western but not yet modern, will be plagued by instability for decades to come.

For these nations we can hope that their progress will be humane and decent; hope that they avoid the excesses of violence which accompanied the development of so many nations of the West. And we should offer to their effort such assistance as we can, or what will be effective.

We cannot continue, as we too often have done in the past, to automatically identify the United States with the preservation of a particular internal order within those countries, or confuse our own national interest with the rule of a particular faction within them. Of course, those in power in these countries will often seek to preserve their position by requesting our help.

Faced with such requests, we must make calm and discriminating judgments as to which governments can and should be helped, which are moving effectively to defend themselves and meet the needs of their people.

Where the central interests of the United States are not directly threatened, I could propose a simple functional test: We should give no more assistance to a government against any internal threat than that government is capable of using itself, through its agencies and instruments. We can help them but we cannot again try to do their jobs for them.

That limitation has, in fact, been forced to our attention. For one thing, Vietnam has proven that all the might and power of America cannot provide or create a substitute for another government, or for the will of another people. But let us understand the full significance of the limitation I suggest. It does not prevent us from aiding any nation against truly external aggression. It does not prevent us from extending reasonable assistance to developing nations. It does prevent us from taking over an internal struggle from a minority government or a government too ineffective or corrupt to gain the support of its own people. It would allow the future of each country to be settled, essentially, by the people of that country . . .

Even within the nations once ruled by Moscow, the forces of national independence and personal freedom are steadily eroding the Soviets' once-unquestioned position. That force can be our strongest ally in the world, if we respect and honor it. It can also be our nemesis, if we continue to ignore it.

The worst thing we could do would be to take as our mission the suppression of disorder or of internal upheaval everywhere it appears. This is even more true if the means for this policeman's role is to be the indiscriminate introduction of American troops into the internal struggles of other nations.

Their presence can transform a factional struggle within a country into a nationalist struggle against foreign domination. Their introduction commits our prestige to the outcome of diverse struggles we may barely understand. It may lead the government and the people to refuse essential sacrifices . . .

We must also keep this entire issue in its proper perspective. For whatever American interests are involved in the internal affairs of the nations of the Third World, they are very slight compared to the substantial threats which potentially lie ahead. These have not been conquered by Vietnam; indeed, in many respects, [it has] made them far worse.

What are they, and how are they met?

One is the challenge of the Soviet Union . . . not a world movement, but itself the second most powerful nation in the world [that] remains for us in many ways a potential threat of unknown magnitude.

Our response, however, is neither unknown nor complex. It is to seek peace, while remaining alert to the danger of mutual destruction posed by our great nuclear capabilities . . .

The second great danger ahead of us is China: one quarter of the world's people, filled with resentment and distrust of the foreigner, and a determination to avenge a century of humiliation . . . For the foreseeable future, China can damage us only to the extent we make it possible—by involving ourselves in a land war with her on the continent of Asia . . .

The third great danger is not from an external enemy. It is from ourselves. This is precisely the danger that, in seeking universal peace, needlessly fearful of change and disorder, we will in fact embroil ourselves and the world in a whole series of Vietnams. But that danger can only be of our own making. The way to avoid overcommitment in rhetoric and action is above all to recognize that America's interest is not in automatic use of military force in the attempt to preserve things as they are elsewhere in the world. Ours is a great and powerful nation. It will continue to be, if we respect the true sources of our own strength and do not spend our men and our resources in the defense of bankrupt regimes on every continent.

For the fourth danger is here at home. It is the danger that absorption in the problems of others will cause us to neglect the health and quality of our own society. We cannot continue to deny and postpone the demands of our own people while spending billions in the name of the freedom of others. No nation can exert greater power of influence in the world than it can exercise over the streets of its own capital. A nation torn by injustice and violence, its streets patrolled by army units—if this is to be our country, we can doubt how long others will look to us for leadership, or seek our participation in their common ventures. America was a great force in the world, with immense prestige, long before we became a great military power. That power has come to us and we cannot renounce it, but neither can we afford to forget that the real constructive force in the world comes not from bombs but from imaginative ideas, warm sympathies, and a generous spirit. These are qualities that cannot be manufactured by specialists in public relations. They are the natural qualities of a people pursuing decency and human dignity in its own undertakings without arrogance or hostility or delusions of superiority toward others; a people whose ideals for others are firmly rooted in the realities of the society we have built for ourselves.

The audience of nearly four thousand students interrupted Kennedy sixteen times with applause. Equally favorable was the press coverage; typical was the reaction of David Broder of the *Washington Post,* who characterized the Bloomington speech as Kennedy's seminal foreign policy address of the campaign. Kennedy would give it again, on a number of occasions and with few modifications.[5]

Crime in America

VINCENNES, INDIANA

APRIL 22, 1968

INDIANAPOLIS, INDIANA

APRIL 26, 1968

Kennedy's call for reappraising America's role abroad found a domestic echo in his views on the topic that opinion polls consistently spotlighted as voters' greatest concern in 1968: crime. Kennedy possessed a personal toughness allied with law enforcement credentials that many white Americans found comforting. Indeed, Kennedy began the 1968 campaign burdened with an image of *excessive* toughness, which had its roots in his aggressiveness as JFK's campaign manager and as chief counsel to the Senate Rackets Committee—a reputation for what some called "ruthlessness," which grew during his relentless pursuit of corrupt union leaders, particularly Teamsters boss Jimmy Hoffa.

Throughout the 1968 campaign, Kennedy made frequent reference to having been the nation's "top cop," and his record as a tough crime-fighter while attorney general gave him some political cover as he continued to ask for compassion, understanding, and aid to poor

Americans, whom the more-comfortable majority was regarding increasingly with suspicion and (especially after the April riots) fear.

The Indiana primary presented a significant political dilemma in this regard: The state was predominantly conservative and overwhelmingly white. Some members of the press corps accused Kennedy of tailoring his message to emphasize his crime-fighting credentials with white audiences in Indiana, but a review of the record bears out the consistency of his message. Journalist Jack Newfield remembered one typical day, during a motorcade

> *through the racially divided and tense steel town of Gary. The black mayor, Richard Hatcher, was balanced on one side of Kennedy, and Tony Zale, the Slavic warrior who came out of Gary's blast furnaces to twice win the middleweight boxing championship, was braced on the other side of the candidate. The open cars rode through the white part of Gary, and then the black part, and Kennedy said precisely the same thing to both races: Jobs were better than welfare, because welfare created dependency and work conferred self-respect; we had to*

be tough on crime; riots were no solution to the problems . . . The reaction was equally enthusiastic in each half of the city.

Crime, however, was the principal topic when Kennedy addressed the Marion County Democratic Committee luncheon in Indianapolis less than two weeks before the May 7 primary. The speech contained detailed suggestions for enhancing public safety and assisting law enforcement, which were rooted in proposals he had been advancing in the Senate since early 1966. The campaign's nearly thirty-page white paper on criminal justice was widely circulated and also excerpted in the May 31, 1968, urban policy release. But Kennedy made clear that crime was a shared American problem that destroyed the sense of a harmonious community that he was striving to create—it was not a "wedge" issue to be used cynically to divide voters.

Crime is an issue that is difficult and dangerous; easily susceptible to illusory promises and false programs; an issue which threatens to divert us from the road to a better nation into blind alleys of suspicion and mistrust.

So let us examine not just the danger of crime but what we can do together to meet the dilemmas of lawlessness . . .

The real threat of crime is what it does to ourselves and our communities. No nation hiding behind locked doors is free, for it is imprisoned by its own fear. No nation whose citizens fear to walk their own streets is healthy, for in isolation lies the poisoning of public participation. A nation which surrenders to crime—whether by indifference or by heavy-handed repression—is a society which has resigned itself to failure. Yet disturbingly, many Americans seem to regard crime as a pervasive enemy that cannot be defeated . . .

Thus the fight against crime is in the last analysis the same as the fight for equal opportunity, or the battle against hunger and deprivation, or the struggle to prevent the pollution of our air and water. It is a fight to preserve that quality of community which is at the root of our greatness; a fight to preserve confidence in ourselves and our fellow citizens; a battle for the quality of our lives.

There were times, however, when a white audience's complacency toward impoverished conditions, which bred more problems than criminality, provoked Kennedy to risk whatever political capital his crime-fighting record gained him. Such an instance came during a critical blitz of southern Indiana, home of the state's most tradition-minded voters, four days before he addressed the Marion County Democrats. On April 22, Kennedy and a phalanx of reporters traveled to the Vincennes Ramada Inn, where the candidate spoke on "The Role of Private Enterprise" to a luncheon of the Civitans, members of the local men's service clubs.

Thomas Congdon Jr., an editor of *The Saturday Evening Post* who was part of the press entourage that day, remembered that

> *Kennedy was late, and the club members had started their meal, and they continued it as he spoke, their eyes on the Salisbury steak instead of the candidate. He gave a version of his standard speech, but he was strained and formal. There were questions about his views on gun control and the Pueblo incident and "God's time"—federally imposed daylight saving time had infuriated Indiana farmers . . .*
>
> *Then, turning a vague question to his own purpose, he gave up all notion of speaking to the audience on the matters that concerned it most—*

such as law and order—and instead tried to stir it with the matters that most concerned him. To the club members—big, heavy men, most of them, well-fleshed and still occupied in shoveling in their lunch—the senator from New York spoke of children starving, of "American children, starving in America." It was reverse demagoguery—he was telling them precisely the opposite of what they wanted to hear.

"Do you know," he asked, voice rising, "there are more rats than people in New York City?"

Now this struck the club members as an apt metaphor for what they had always believed about New York City, and a number of them guffawed.

Kennedy went grim and, with terrible deliberateness, said, "Don't . . . laugh . . ."

The room hushed and the program soon broke up, with the audience still ruffled and confused.

Despite that confrontational spirit, or perhaps because his peculiar fierce compassion spoke to hardscrabble farmers as much as poor black mothers, Kennedy won the Indiana primary on May 7. He captured two thirds of the counties in Civitan country (southern Indiana), and he swept all seven of the counties that had been George Wallace bastions four years before. It was his first electoral challenge to McCarthy and the Johnson administration, and Kennedy won despite concerted attacks by the Pulliam newspaper chain (Indiana's largest), beating Governor Roger D. Branigin (who led a strong Democratic machine as Vice President Hubert Humphrey's stalking horse) and McCarthy and his legion of youthful campaign workers. Kennedy triumphed despite a demographic and ideological mix diametrically different from his normal support base—and yet he did so without softening his moral crusade.

A week later Kennedy won by an even larger margin in Nebraska. There, not facing a Branigin-type, favorite-son candidacy, which cut into his vote totals in Indiana, Kennedy swept every sector of Nebraska. He had particularly strong support among blacks (a minuscule portion of the electorate) and, tracking his performance in Indiana, among white ethnics and farmers. Also characteristically, the candidate was the campaign's best asset, winning twenty-four of the twenty-five Nebraska counties in which he campaigned (losing the lone McCarthy stronghold, which included the University of Nebraska, by two votes). Nebraska was a revelation for many of the traveling press and for some of Kennedy's closest aides. One of them, Jeff Greenfield, later reflected that

Nebraska was the first time that I really saw . . . Robert Kennedy's ability to command very different kinds of political constituencies. In Nebraska I saw the way he related to people who had nothing in common with him at all. You don't think of Robert Kennedy as a man of the cities, exactly, but he was of New York, Boston, and Washington—that scene. And he certainly wasn't a farmer. Yet, there was a kind of communication between him and . . . almost Grant Wood kind of characters in a sense—leather-skinned, very hard-working people, very traditional values . . . they probably didn't like . . . antiwar demonstrators. Outside of Omaha, they might not have seen a Negro in their lives . . . the people who would be the last people in the world you would imagine Robert Kennedy to have any relationship with, who really had come not to scream and to yell and to cheer, but to listen. And he got through . . . It really taught me a lot about him.

Reforming the Welfare System

PRESS RELEASE; LOS ANGELES, CALIFORNIA

MAY 19, 1968

The primary pace quickened after Nebraska: Oregon's was two weeks later and the season's second-richest delegate prize, California, one week thereafter. Kennedy focused on the latter, racing through a hectic three-day schedule in the Golden State.

As the foray was beginning, the campaign released a major white paper setting forth Kennedy's views on welfare reform. (He had outlined its themes to a crowd of ten thousand at Los Angeles Valley College on May 15.)[6] Kennedy's insistence on real and fundamental change had never been satisfied merely by the passage of legislation or the establishment of programs. Increasingly after entering the Senate, he strongly questioned the efficacy of many government actions designed to benefit poor Americans—including those programs he had helped design. Key among the failures, in Kennedy's view, was the welfare system, which he had criticized for years. His approach transcended traditional ideology—for welfare failed those who funded it as well as those who received its assistance.

Perhaps the area of our greatest domestic failure is in the system of welfare—public assistance to those in need. There is a deep sense of dissatisfaction, among recipient and government alike, about what welfare has become over the last thirty years, and where it seems to be going.

Welfare is many things to many people. To the recipient it may be the difference between life and starvation, between a house and homelessness, between the cold wind and a child's coat. To the taxpayer—facing inflation in the cost of living, paying for his home and educating his children—welfare may be an unwarranted imposition on an already overburdened tax bill. To certain politicians, willing to oversimplify and confuse the issue, it may be a means to easy popularity . . .

The bill is rising further every day.

With all this enormous expenditure, might we not expect that the recipient would be satisfied? Yet the fact is that they are not. They are as dissatisfied with the welfare system as is anyone in the U.S.

. . . Is this rank ingratitude—or is it an indication of how the welfare system has failed? For what are we to make of a system which seems to satisfy neither giver nor recipient—which embitters all those who come in contact with it?

The worst problem is in our very concept of welfare . . . Welfare began as a necessary program of assistance for those unable to work. But we have tried as well to make it the easy answer to the complex, but by no means insurmountable, problem of unemployment . . .

[The unemployed] are men like other men. They marry and have children; or they do not marry but have children just the same. In either case, they often leave home under the strain of joblessness and poverty. We have dealt with the resulting female-headed families not by putting the men to work but by giving the mothers and children welfare. They might have wanted fathers and husbands; we have given them checks. In fact, the welfare system itself has created many of these fatherless families—by requiring the absence of a father as a condition for receiving aid; no one will ever know how many left their families to let them qualify for assistance so that they might eat, or find a place to live.

More basically, welfare itself has done much to divide our people, to alienate us one from the other. Partly this separation comes from the understandable resentment of the taxpayer, helplessly watching your welfare rolls and your property tax rise. But there is greater resentment among the poor, the recipient of our charity. Some of it comes from the brutality of the welfare system itself: from the prying bureaucrat, an all-powerful administrator deciding at his desk who is deserving and who is not, who shall live another month and who may starve next week.

But the root problem is in the fact of dependency and uselessness itself. Unemployment means having nothing to do—which means nothing to do with the rest of us. To be without work, to be without use to one's fellow citizens, is to be in truth the Invisible Man of whom Ralph Ellison wrote . . .

We often quote Lincoln's warning that America could not survive half slave and half free. Nor can it survive while millions of our people are slaves to dependency and poverty, waiting on the favor of their fellow citizens to write them checks. Fellowship, community, shared patriotism—these essential values of our civilization do not come from just buying and consuming goods together. They come from a shared sense of individual independence and personal effort.

They come from working together to build a country—that is the answer to the welfare crisis.

The answer to the welfare crisis is work, jobs, self-sufficiency, and family integrity; not a massive new extension of welfare; not a great new outpouring of guidance counselors to give the poor more advice. We need jobs, dignified employment at decent pay; the kind of employment that lets a man say to his community, to his family, to his country, and most important, to himself, "I helped to build this country. I am a participant in its great public ventures. I am a man." . . .

It is a myth that all the problems of poverty can be solved by ultimate extension of the welfare system to guarantee to all, regardless of their circumstances, a certain income paid for by the federal government. Any such scheme, taken alone, simply cannot provide the sense of self-sufficiency, of participation in the life of the community, that is essential for citizens of a democracy . . .

Certainly, all the proposals for various systems for income maintenance deserve careful study. But if there is anything we have learned in the last three years, it is that we cannot do everything at once—that we must understand, establish, and adhere to a clear sense of national priorities. The priority here is jobs. To give priority to income would be to admit defeat on the critical battlefront . . .

Work is a mundane and unglamorous word. Yet it is, in a real sense, the meaning of what the country is all about—for those of us who live in affluent suburbs and for our children no less than for the children in the ghetto. Human beings need a purpose. We need it as individuals; we need to sense it in our fellow citizens; and we need it as a society and as a people.

In the sections of the press release omitted above, Kennedy called for replacing the welfare bureaucracy with an automatic system based entirely on need, having national standards and substantial incentives (not penalties) for people to work and raise happy families. The speech included one of the earliest calls for improved and expanded day care and a passionate argument against a system that penalized out-of-work fathers who remained with their families. Many of the inadequacies he spotlighted have remained in the welfare schema in the decades after his death.

Forging a New Politics

SAN FRANCISCO, CALIFORNIA
MAY 21, 1968

Two days after his welfare proposal was released, Kennedy addressed journalists at a luncheon in San Francisco. He returned to a frequent theme, so much a part of his Senate career and one also described for a far different, rural audience in Scottsbluff, Nebraska, a month before: a new view of politics that could address the nation's concerns, mobilize its citizens, and reflect in its foreign policy its deepest beliefs about human nature. His emphasis on jobs, the linchpin of his welfare reform, was equally insistent here.

Kennedy was, in David Halberstam's phrase, a transitional figure in a transitional year. His career began in the time of the old politics, and his iron-handed management of his brother's 1960 presidential campaign had proven his mastery of the genre. But the old relationships and the familiar shibboleths no longer rang true by 1968, and Kennedy paid the political price for delaying his transition: His personal caution, and the advice of the old school, had turned him away from challenging Johnson for the nomination throughout 1967 and the early months of 1968. This allowed Eugene McCarthy to co-opt student support and spearhead the antiwar movement, and Kennedy's tardy entrance exacerbated his reputation for ruthless ambition.

In fact, Kennedy had presented an amalgam of pragmatism and moral urgency throughout his entire career, particularly in the Senate. A lifelong advocate of focused governmental action, Kennedy nonetheless maintained what Walinsky characterized as "a very strong distrust of the bureaucratic process and . . . government." The year-long struggle to fashion a revitalization effort in Bedford-Stuyvesant reaffirmed for Kennedy the importance of creating institutions in which individuals and communities could take a direct role. He believed that he could inspire such involvement by evoking communal efforts that had provided signal triumphs throughout America's history, and by offering a social philosophy that balanced responsibility and opportunity, allowing the majority to participate in (and feel comfortable with, and benefit from) national efforts principally designed to assist racial and other minorities.

Kennedy wove these concerns and his view of a world role for America that was neither interventionist nor isolationist (as the positions had become defined) into his remarks for the San Francisco press gathering on May 21, a week before the Oregon primary and two weeks before California's.

It is clear by now that 1968 will go down as the year the new politics of the next decade or more began. It is the year when the existing political wisdom had proven unable to cope with the turbulence of our times, inspire our young people, or provide answers to problems we face as a nation. And therefore this is the year when the old politics must be a thing of the past.

But if this is true—and I profoundly believe that it is—then there is no more important question than what the new politics is. What are its components, and what does it mean to the future of the country?

The most obvious element of the new politics is the politics of citizen participation, of personal involvement . . .

The [next] priority for change . . . is in our policy toward the world. Too much and for too long, we have acted as if our great military might and wealth could bring about an American solution to every world problem . . .

We must be willing to recognize that the world is changing, and that our greatest potential ally is the simple and enduring fact that men want to be free and independent . . .

The second demand of the new politics is here at

home. It begins with the recognition that federal spending will not solve all our problems, and that money cannot buy dignity, self-respect, or fellow feeling between citizens.

The new politics will recognize that these things do come from working together to build a country; and it will make the first domestic task of the next administration the creation of dignified jobs, at decent pay, for all those who can and want to work. It will be for a far better public assistance system, one which affords adequate help to those who cannot work, without the indignities and random cruelties which afflict the present welfare systems.

But the first priority will be jobs—a share in our great common enterprises, a life as men—for all those who now linger in idleness, unseen, unheard, and unwanted.

The third element of the new politics is to halt and reverse the growing accumulation of power and authority in the central government in Washington, and to return that power of decision to the American people in their own local communities. For the truth is that with all the good that has been accomplished over the last thirty years—by unemployment compensation, Medicare, and fair labor standards, by the programs for education, housing, and community development—for all that, still the truth is that too often the programs have been close to failures.

If this is to change—and it must change—we must recognize that the answer is not just another federal program, another department or administration, another layer of bureaucracy in Washington. The real answer is in the full involvement of the private enterprise system—in the creation of jobs, the building of housing, the provision of services, in training and education and health care.

Through a flexible and comprehensive system of tax incentives, we can and should encourage private enterprise to devote its energies and resources to these great social tasks, which I believe [it] could help us to accomplish with far less cost, far more effectiveness, and far more freedom than with more government programs . . .

The new politics of 1968 has a final need: and that is an end to some of the clichés and stereotypes of past political rhetoric. In too much of our political dialogue, "liberals" have been those who wanted to spend more money, while "conservatives" have been those who wanted to pretend that all problems should solve themselves. Emerson once wrote that "conservatism makes no poetry, breathes no prayer, has no invention; it is all memory," while reform, he wrote, "has no gratitude, no prudence, no husbandry."

But the times are too difficult, our needs are too great, for such restricted visions. There is nothing "liberal" about a constant expansion of the federal government, stripping citizens of their public power—the right to share in the government of affairs—that was the founding purpose of this nation. There is nothing "conservative" about standing idle while millions of fellow citizens lose their lives and their hopes, while their frustration turns to fury that tears the fabric of society and freedom.

What we do need . . . is a better liberalism and a better conservatism. We need a liberalism, in its wish to do good, that yet recognizes the limits to rhetoric and American power abroad; that knows the answer to all problems is not spending money.

. . . We need a conservatism, in its wish to preserve the enduring values of the American society, that yet recognizes the urgent need to bring opportunity to all citizens, that is willing to take action to meet the needs of the future.

What the new politics is, in the last analysis, is a reaffirmation of the best within the great political traditions of our nation: compassion for those who suffer, determination to right the wrongs within our nation, and a willingness to think and to act anew, free from old concepts and false illusions.

That is the kind of politics—that is the kind of leadership—the American people want.

Oakland community adviser the Reverend Hector Lopez believed that, particularly for blacks and other minorities, Kennedy's amalgam of new politics and his demonstrated passion and sincerity provided an extraordinary appeal: "What can you call Bobby? The last of the great liberals? He wouldn't have liked that. I know he wouldn't. I guess I'd have to say he was the last of the great believables."

In Kennedy's 1968 campaign, philosophy and political strategy were served by the same tactics. Neither his foreign nor domestic policy views fit neatly into the prevailing ideological camps, allowing a constituency encompassing urban blacks, ethnic blue-collar whites, and farmers. And as the insurgent candidate, Kennedy was buoyed by growing popular revulsion with politics as usual. Peter Edelman recalled that

Kennedy used to say that himself often, privately: that people are just sick of politicians. And they

are looking . . . for just an honest man. This particularly became a search and a quest when Lyndon Johnson was in the White House and was such a liar . . . with the consequences of killing thousands of Americans . . . The new politics . . . [was] the consequence of the fact that the American people got tired of being lied to. So they wanted to be talked to directly. They didn't want to be talked to in terms of "on the one hand" and "on the other hand." They just wanted to hear somebody who was directly responsive to their concerns. So that the style had to change. You didn't deal for the votes of people with bosses because they wouldn't stand for that anymore . . . In addition to that you were trying . . . to unseat an incumbent president, up to a certain point, and after that to overturn an established machine, which was all going to another candidate, the vice president . . . The only way you could do that was to go over the heads of the [power brokers] . . . to the people directly and through the media in as many primary states as possible and then in as many . . . other forms of challenging the established way of doing things as possible. And in turn to demonstrate to those delegates who were hand-picked or to their manipulators . . .

Having done that, that the only logical and rational course was to support you . . . because in their cold self-interest, it would help elect their local candidates much better than a Hubert Humphrey would.

Remarks on Gun Control

ROSEBURG, OREGON

MAY 27, 1968

Kennedy's approach, successful in head-to-head contests with McCarthy in Indiana and Nebraska, had no traction in Oregon. Kennedy's message never connected with voters in a state the candidate privately described as one giant suburb; as Kennedy told one reporter, "Let's face it; I appeal best to people who have problems." McCarthy had pulled out of Nebraska early and committed his best organizers to Oregon and California, and their efforts were matched by phone banks and similar support from friends of the administration within the AFL-CIO official structure, aiming to embarrass the candidate they felt would be Vice President Humphrey's only viable challenger.

Gun control became a flash-point issue in Oregon. Co-sponsor of legislation regulating firearms,[7] with his brother's death by a mail-order gun a constant reminder, Kennedy did not duck the topic. The day before the primary, he visited Roseburg, Oregon, in the state's southwest corner, where timbering drove the economy and recreational hunting was widespread. Warned by the local sheriff that he'd face hostility, Kennedy's face was "grim" (according to the traveling *New York Times* reporter)[8] as he approached the microphone on the steps of the Douglas County courthouse in front of a crowd estimated at fifteen hundred—roughly a tenth of the town's population. Before him were signs reading PROTECT YOUR RIGHT TO KEEP AND BEAR ARMS.

"I see signs about the guns," Senator Kennedy began. "I'm wondering if any of you would like to come and explain."

According to the *New York Times* report,

> *[A] heavyset man in a lumber-jacket stepped forward and said: "I'm Bud Stone. The signs refer to the Senate bill [recently passed] and we think it's a backdoor bill for the registration of guns and it would let the Supreme Court make the final decision."*
>
> *When the candidate got the microphone back, he said he understood that gun legislation was a big issue in this lumber town and that there had been broadcasts on the local radio station opposing the Senate bill.*
>
> *"If we are going to talk about this legislation, let's talk about it honestly and not say that it does something that it does not do."*

All [the legislation] requires is that when somebody purchases a gun through a mail order or you send a gun or a rifle across a state line that you abide by the law of a particular state . . . At the present moment, a person who is insane, a man with a long criminal record of having killed a dozen people, can go in and buy a rifle. Now if you think that that makes sense for all of us . . . A person who is four years old can buy a rifle now . . . A man on death row in Kansas, who had killed a half a dozen people, and someone there sent for a rifle, through the mail from Chicago, for him to have a rifle while he was waiting on death row, after killing people . . . and the rifle was sent to him.

Now, does that make any sense, that you should put rifles and guns in the hands of people who have long criminal records, or people who are insane or people who are mentally incompetent, or people who are so young they don't know how to handle rifles and guns? I'd just ask you.

I just present the case—it's presented to you, and I know that it's presented on the radio here, and—[*gesturing*] I'm not making any reflections on this gentleman—is presented by the John Birch Society as somebody who is going to come in, the federal government is going to come in and take your guns away and take your rifles away. Nothing is going to happen about that. Anybody can have—just as you can have an automobile. I hear that . . . here in this community, it's described as the way Nazi Germany started. Well you can say that registering an automobile is the way Nazi Germany started.

All we are talking about is having guns not in the hands—anybody can have a gun, anybody can have a rifle—but a person who's got a criminal record or is in an insane institution or is mentally incompetent shouldn't have a rifle or a gun. Is there anybody out here that thinks those people should have rifles or guns?

[*Smattering of applause and* Nos]

That's all the legislation does. It doesn't stop anybody from having a rifle or a gun, so [your sign] PROTECT YOUR RIGHT TO KEEP AND BEAR ARMS is just misleading the American people and is misleading everybody else.

Nobody is going to come in and take your guns away. . . . With all the violence and murder and killings we've had in the United States, I think you will agree that we must keep firearms from people who have no business with guns or rifles.

[*A man in a cowboy hat booed loudly and shouted,*

"They'll get them (guns) anyway," the New York Times *reported.*]

The next day, Eugene McCarthy won the primary by six points, the first time a member of the Kennedy family had suffered a defeat in twenty-seven electoral contests.

The day after the primary, at a press conference in California, Kennedy read a prepared statement that said "these results represent a setback to my prospects for receiving the presidential nomination of my party, a setback, as I have previously stated, which I could ill afford." Characteristically, in response to a question, he took complete responsibility: "I lost because I didn't do well enough. The only fault is me." Kennedy made a point of sending an early and gracious congratulatory telegram to McCarthy, who had not extended a similar courtesy after Kennedy's victories in Indiana and Nebraska.

California Victory Speech

LOS ANGELES, CALIFORNIA

JUNE 4, 1968

June 4 was primary day in both California and South Dakota. Although he had campaigned in the latter, Kennedy had focused most of his resources, including his own time, on the delegate-rich Golden State, which stood only behind New York as the biggest preconvention prize.

McCarthy had tried to capitalize on his upset victory in Oregon. There, he had made much of Kennedy's refusal to debate him. Kennedy, having conceded that his defeat had changed the political dynamic, agreed to a televised debate in California on June 1. After such a buildup, the actual event was judged as mild by most observers, but the pundits felt Kennedy had done the most to improve his position.[9]

In part, he had outperformed expectations, and in part he simply had worked harder in preparation, and it showed. The same unflagging discipline marked the last day of campaigning: In twelve hours he traveled from Los Angeles, to San Francisco, to Long Beach, via motorcade through Watts, to San Diego, and ultimately back to Los Angeles. His extreme exhaustion (and a probable stomach virus) left him sick and somewhat dazed in his final rally at San Diego's El Cortez Hotel, where the crowd was so large that it had to be divided into two seatings, for the candidate to address in sequence.

Primary day was typical for southern California in June: cool, under a fog-gray and sullen sky. Kennedy slept late and spent the day with his wife and some of his children at a friend's Malibu beach house. Reports from his campaign aides made it clear that at least one organizational imperative was being met: Turnout in minority communities was immense, and the vote for Kennedy in some precincts broke 90 percent. Largely through the concerted effort of the United Farm Workers (under Cesar Chavez's direct supervision), Mexican Americans voted for Kennedy with nearly absolute loyalty—Jack Newfield's analysis showed fourteen of every fifteen voted for the only candidate to have made the farm workers' cause his own.

California was not the only state in play on June 4. In South Dakota, the political goal required the loyalty of the farm owners, more than the farm hands, and the situation was complicated by the appearance on the ballot of two neighbors: Minnesota's Humphrey (a South Dakota native) and McCarthy. Kennedy's ally, South

Dakota Senator George McGovern, called the candidate with the state's results: Kennedy had polled more than his two opponents combined.

It took slightly longer for the California totals to be tabulated (the evening marked Los Angeles County's first use of computer punch cards as ballots), but shortly before midnight, Kennedy came down from his campaign suite in Los Angeles's Ambassador Hotel to thank his supporters and claim his twin victories (he would beat McCarthy in California by 4.5 percent) in extemporaneous remarks.

I am very grateful for the votes that I received—that all of you worked for . . . in the agricultural areas of the state, as well as in the cities . . . as well as in the suburbs. I think it indicates quite clearly what we can do here in the United States. The vote here in the state of California, the vote in the state of South Dakota: Here is the most urban state of any of the states of our union, and South Dakota, the most rural state of any of the states of our union. We were able to win them both.

The candidate was interrupted by applause and shouts of "Bobby Power" and "Kennedy Power."

I think we can end the divisions within the United States. What I think is quite clear is that we can work together in the last analysis. And that what has been going on with the United States over the period of the last three years—the divisions, the violence, the disenchantment with our society, the divisions, whether it's between blacks and whites, between the poor and the more affluent, or between age groups, or over the war in Vietnam—that we can start to work together again. We are a great country, an unselfish country, and a compassionate country. And I intend to make that my basis for running over the period of the next few months . . .

So my thanks to all of you, and it's on to Chicago, and let's win there.

Exiting through a crowded hotel kitchen to avoid the even deeper crush of the ballroom, Kennedy was shot in the head by an assassin.

Doctors at Good Samaritan Hospital labored throughout June 5, to no avail, and in the end, only Kennedy's indomitable heart kept up the struggle. He died at 1:44 A.M. on June 6, 1968. Robert Kennedy was forty-two.

Appendix
Robert F. Kennedy Human Rights

Led by human rights activist and lawyer Kerry Kennedy, Robert F. Kennedy Human Rights (http://rfkhumanrights.org) has advocated for a more just and peaceful world since 1968. It works alongside local activists to ensure lasting positive change in governments and corporations. The team includes prominent attorneys, advocates, and leaders in business and government united by a commitment to social justice.

Whether in the United States or abroad, the programs have pursued justice through strategic litigation on key human rights issues, educated millions of children in human rights advocacy, and fostered a social good approach to business and investment.

Chronology
Robert Francis Kennedy

November 20, 1925	Born in Brookline, Massachusetts
March 1938	Sails to England where father is U.S. ambassador
October 1943	Enlists in the Naval Reserve
August 1944	Oldest brother (Joe) dies on a volunteer bombing mission over occupied Europe
May 30, 1946	Honorable discharge, after having served on destroyer USS *Joseph P. Kennedy Jr.*

September 1946	Enters Harvard University as a junior; earns two varsity football letters
November 1946	Brother John wins U.S. congressional seat for Massachusetts's Eleventh District
March–June 1948	After graduation from Harvard, travels through Europe and the Middle East
September 1948	Enters University of Virginia School of Law
June 17, 1950	Marries Ethel Skakel
June–November 1952	Manages John's successful U.S. Senate campaign
January–July 1953	Assistant Counsel for the Senate Permanent Subcommittee on Investigations (Chair: Joseph McCarthy)
February 1954	Becomes Chief Counsel for Democratic minority on same subcommittee

January 1955	Chief Counsel for majority when Democrats win control of the Senate
1957–1959	Chief Counsel for the Senate Select Committee on Improper Activities in Labor and Management (McClellan, or Rackets, Committee)—Teamster hearings
1959–1960	Manages John's presidential campaign
1960	*The Enemy Within* published (about the labor investigations)
January 20, 1961	Sworn in as U.S. attorney general
May 21, 1961	Orders U.S. marshals to Montgomery, Alabama, to protect Freedom Riders
October 1, 1962	James Meredith becomes first black enrollee at the University of Mississippi after federal troops quell campus riots

October 16–28, 1962	Cuban Missile Crisis
November 22, 1963	President Kennedy assassinated
August 22, 1964	Declares Senate candidacy
September 3, 1964	Resigns as attorney general
January 1965–June 1968	U.S. senator from New York
March 16, 1968	Declares candidacy for president
April 4, 1968	Dr. Martin Luther King Jr. assassinated
May 7, 1968	Wins Indiana primary
May 14, 1968	Wins Nebraska primary
May 28, 1968	Loses Oregon primary
June 4, 1968	Wins California and South Dakota primaries; shot leaving victory speech

Notes

PART ONE: JOURNALIST, SENATE COMMITTEE COUNSEL, CAMPAIGN MANAGER

1. The Battle Act, named for its sponsor, Representative Laurie C. Battle (Democrat, Alabama) and signed into law by President Truman on October 26, 1951, was designed to cut off U.S. aid to any nation that sent arms, military equipment, or other strategic materials to the Soviet bloc.
2. Hoffa unceremoniously ousted Beck, whom the hearings had exposed as a thief and a paper lion, when the Teamsters met in August 1957, and he took over the union in October that year, in an election that Kennedy said was "rigged from start to finish."

PART TWO: MR. ATTORNEY GENERAL

1. In 1959 Prince Edward County's all-white board of supervisors, refusing to obey a federal court desegregation order, closed the public schools, denying education to 1,500 black children while white children went to "private schools" supported mostly with public funds. In 1961 the Justice Department moved in federal court for an order forcing the county to reopen the schools.
2. In 1957 President Eisenhower had to send paratroopers to Little Rock to prevent angry whites from blocking court-ordered desegregation of Central High School. In 1961 the Louisiana legislature was threatening to withdraw school funding if a federal judge's order desegregating some New Orleans schools was enforced. The Justice Department was moving to enforce the order. The schools were desegregated peacefully the following September.
3. Charlayne Hunter-Gault later became an accomplished journalist at the *New York Times*, *The New Yorker*, NPR, and on then–*The MacNeil/Lehrer Report*, which became *PBS NewsHour.*

PART THREE: THE SENATE YEARS

1. The General Aniline and Film Corporations (GAFC) case was one of the most contentious and complex

corporate lawsuits of the era. Stock in the company was seized by the United States in 1942, because the company was ultimately owned by IG Farben, the German chemical cartel that had been vital to the Nazis. A Swiss firm sued to recover the stock; in response, the U.S. government alleged the plaintiff was a front for IG Farben. The case wound its way up and down the judicial system, and Congress tried unsuccessfully for years to set up a mechanism whereby the federal government would auction its interest in GAFC and escrow the proceeds until the underlying stock ownership could be adjudicated. Finally, in 1962, Congress unexpectedly solved the issue, authorizing the sale of GAFC, with the proceeds going into the War Claims Fund, also established under the legislation—a $500 million program to discharge American claims for World War II losses caused by German entities. As attorney general, Kennedy was responsible for the government's share ownership and had to oversee the complicated sale and settlement of the underlying litigation. Kenneth Keating falsely branded the settlement as Kennedy's initiative and alleged it was favorable to Nazi collaborators. More details are available online at http://www.nytimes.com/1964/04/19/progress-made-in-aniline-case.html.

2. Beckwith had been twice tried in 1964 for the previous year's sniper slaying of Medgar Evers, the Mississippi field secretary of the National Association for the Advancement of Colored People. On both occasions, all-white juries were unable to reach a verdict, and the charges were dismissed in 1969. In 1990, after new evidence of Beckwith's guilt surfaced, a grand jury indicted him again, and on August 4, 1992, a judge in Jackson, Mississippi, refused Beckwith's request to be freed because of deteriorating health and memory, and ordered him to stand trial. He was convicted in 1994 and died in 2001.
3. The incident is recounted in Gerald Gardner's *Robert Kennedy in New York* (New York: Random House, 1965), p. 72.
4. Aide Adam Walinsky modified this idea after Kennedy's death into a visionary proposal for a Police Corps, trading college scholarships for the commitment to serve in law enforcement. For two decades, Walinsky single-handedly lobbied Congress, wrote op-ed articles, and pushed for his program. In 1991, creation of the Police Corps became part of an omnibus crime bill; it survived a House-Senate conference committee, but the bill was blocked on other issues by Senate Republicans. Finally created by President

Clinton, the Corps fell prey to Republican budget cutting in the early 2000s.

5. Johnson had declared an "unconditional war against poverty" in his first State of the Union address in January 1964, and on March 16 of that year he requested nearly $1 billion in federal funds to wage the effort. Johnson said it was a "war . . . not only to relieve the symptoms of poverty, but to cure it, and, above all, to eliminate it." It would be directed by the newly created Office of Economic Opportunity. Some of the programmatic elements of the War on Poverty became part of Johnson's vision of the "Great Society," which he described in a May 22, 1964, speech and then expanded upon in his second State of the Union address on January 4, 1965. Johnson's domestic policy aims were the most ambitious since Franklin Roosevelt's New Deal. Included within Johnson's broad vision for the Great Society were massive federal initiatives in the areas of urban renewal, health care, education, poverty, and hunger. Some programs, such as Head Start, continue today and enjoy widespread support; others were found ineffective, and still other initiatives floundered without sufficient funding.
6. Marian Wright married Kennedy aide Peter Edelman fifteen months after the two met while pre-

paring for the April 1967 hearings. After years of leadership in civil rights, she became convinced that the greatest needs were of the young, and in 1973 she formed the nation's first advocacy organization for children, the Children's Defense Fund. The organization is renowned for its lobbying and public information efforts, and for the depth and probity of its statistical analyses of the conditions of America's children.

7. From Kotz's book *Let Them Eat Promises: The Politics of Hunger in America,* which grew out of his coverage of Kennedy in connection with these 1967 hearings. After Kennedy's death, Kotz told Peter Edelman that "what he was going to do to contribute to the perpetuation of [Kennedy's] memory was to finish that book."
8. Some months later, Kennedy used a chance meeting at a party with Don Hewitt, the head of CBS News (and later executive producer of *60 Minutes),* to emphasize the importance of television in educating the country about the hungry in their midst. Stimulated by that conversation, Hewitt authorized a documentary, *Hunger in America,* which shocked the nation when it aired in April 1968, just as Kennedy was beginning his run for the presidency.

9. Hippolyte Taine was a nineteenth-century French critic and historian whose belief that man was the product of heredity and environment became the central tenet of the naturalistic school.
10. *Alianza para el Progreso,* the Alliance for Progress, was President Kennedy's moniker for a new relationship he intended to create between the United States and Latin America. First announced in a speech in Tampa, Florida, in the last month of the 1960 presidential campaign, the Alliance was, in John Kennedy's words, "a great common effort to develop the resources of the entire hemisphere, strengthen the forces of democracy, and widen the vocational and educational opportunities of every person in the Americas." Kennedy also promised "unequivocal support to democracy," economic assistance, and encouragement of internal programs such as land reform. The president officially launched the Alliance in a March 13, 1961, White House ceremony.
11. Allard Lowenstein exerted an extraordinary influence over American politics in the 1960s. Called the "supreme agitator" of his age, Lowenstein began as a protege of Eleanor Roosevelt; he electrified the United Nations as a youth, describing his travels through southern Africa and his opposition to apart-

heid; organized student involvement in the U.S. civil rights struggles in the South; and spearheaded the "Dump Johnson" movement, which, beginning in 1967, coalesced antiwar sentiment, ultimately fueling the 1968 presidential candidacies of Eugene McCarthy and Robert Kennedy, and forcing Lyndon Johnson out of the race. Admired for his passion, eloquence, and integrity by figures as diverse as Kennedy, Martin Luther King Jr., and conservative intellectual William F. Buckley Jr., Lowenstein was shot and killed in 1980 by a deranged former associate.

12. Although the three shotgun blasts left Meredith with seventy-five pellets embedded in his body, he survived and rejoined the marchers for the arrival rally in Jackson.
13. A snip is a small South African bird.
14. Our website (rfkspeeches.com) provides a more detailed telling of the times and the substantive and political objectives of Kennedy's approach to Vietnam as a senator, including all materials on the topic that were included in the original 1993 edition. That depth is more reflective of the historical record and useful for charting his changing views on one of the most critical issues of his era and his broader personal evolution from a traditional cold warrior into a

peace candidate. Even so, the history of Vietnam—even that of only the period since World War II, and even if further focused on matters directly relevant to the U.S. involvement there—is rich and complex, and beyond the scope of this volume. The discussion in this edition aims merely to provide a skeletal background for Robert Kennedy's views, with particular attention to events during President Kennedy's administration, a political and personal legacy crucial to Robert Kennedy's later thinking.

15. Japan controlled Vietnam after colonial power France fell to the Nazis in 1940—first tacitly, via the Vichy French; by 1941, exercising effective control; and for the five months before Japan's surrender in August 1945, directly, having imprisoned Vichy officials.
16. In the early fall of 1963, America had 16,732 military advisers in the country (there had been 685 when John Kennedy took office), and combat fatalities from 1961 through the end of 1963 numbered 73. It appeared to some of his senior advisers that President Kennedy unequivocally intended, after his 1964 reelection, to follow through on the withdrawal plan he had commissioned from Robert McNamara. Both the president and his brother assumed there would be plenty of time to disengage. Events in Dallas on November 22, 1963, proved them wrong.

17. For a stunning exploration of the feud between Kennedy and Johnson, see Jeff Shesol, *Mutual Contempt* (New York: Norton, 1997).
18. The crew was not released until nearly a year later (on December 23, 1968).
19. Congress initially granted Johnson broad powers to prosecute the war after the Tonkin Gulf incident in August 1964, which the White House described as an unprovoked attack by North Vietnamese vessels on the American destroyer *Maddox*. History has revealed facts as far more murky (some doubt that any attack occurred at all), and it is unquestionable that Johnson seized on the incident to ram through Congress the discretionary war powers he wanted. Kennedy feared Johnson's ability to manipulate events (or the public perception of them) for political ends.

PART FOUR: THE 1968 PRESIDENTIAL CAMPAIGN

1. William Allen White (1868–1944) owned and edited the Emporia, Kansas, *Gazette*. An influential exponent of grass-roots Republican views in the early twentieth century, White also authored biographies of Woodrow Wilson and Calvin Coolidge.
2. Quoting William Appleman Williams; the subsequent two sentences paraphrase Sophocles.

3. In 1966 a drifter named Richard Speck fatally stabbed eight student nurses in Chicago. Later that same year, a former Eagle Scout, ex-Marine, and model student, Charles Whitman, killed his wife and mother, then climbed with his rifle to the top of the University of Texas tower and for more than an hour and a half methodically killed strangers in the quadrangle below. Fourteen people died that day before police killed Whitman.
4. Greenfield, who described the incident in his oral history interview for the Kennedy Library, reflected that "when the news of John Kennedy's death first came out, the news reports had the name wrong . . . backwards . . . And that's the way he remembered it, because obviously he never took another look at it again . . . He was thinking of [Medgar] Evers and obviously very clearly thinking of the death of his brother."
5. In fact, as Peter Edelman pointed out, Kennedy frequently "recycled" his substantive speeches on the campaign trail, spicing up the policies developed during his previous three years in the Senate with local statistics, anecdotes, and examples.
6. Walinsky remembered that Kennedy deliberately chose the topic for his campus appearance: "There were two parts to it: One was . . . just overtly po-

litical: you couldn't go on to a campus and compete with McCarthy about the war . . . so you had to talk social justice. The other thing . . . was substantive . . . which was that . . . he just hated going in there and . . . just throwing them red meat . . . he always wanted to challenge them."

7. Kennedy was a co-sponsor of Connecticut Democratic Senator Thomas J. Dodd's SB 1592, putting federal restrictions on mail-order gun sales.
8. *New York Times*, May 28, 1968. This wasn't to be the last time Roseburg and gun violence made national news. On October 1, 2015, a twenty-six-year-old student at Umpqua Community College fatally shot an assistant professor and eight students, and wounded eight others. The killer was wounded in a shoot-out with police and then killed himself. It was the deadliest mass shooting in Oregon's modern history.
9. The greatest difference between the candidates that surfaced that evening was regarding the nation's ghettos. McCarthy wanted to break them up and disperse the residents; Kennedy argued that a fully integrated society was years away and that the immediate need was for reconstruction. His position reflected a fundamental pillar of his urban strategy since his three speeches on the topic in January 1966;

during this debate, his language was more overtly political: "To take these people out, put them in suburbs where they can't afford the housing, where their children can't keep up with the schools, and where they don't have the skills for the jobs, it is just going to be catastrophic . . . You say you are going to take ten thousand black people and move them into Orange County."

Sources

RFK's official papers reside at the John F. Kennedy Presidential Library and Museum in Boston. For Sections Two, Three, and Four, we used the final, official copy of each speech and, wherever available, examined the podium copies for Kennedy's last-minute changes. The following materials were especially helpful for the contextual materials that precede and follow each entry:

PART ONE: JOURNALIST, SENATE COMMITTEE COUNSEL, CAMPAIGN MANAGER

I. Books about Robert Kennedy

Guthman, Edwin O. *We Band of Brothers.* New York: Harper & Row, 1971.

Schlesinger, Arthur M., Jr. *Robert Kennedy and His Times.* Boston: Houghton Mifflin, 1978.

II. Other Sources

Donovan, Robert J. *Conflict and Crisis.* New York: W. W. Norton & Co., 1977.

Kennedy, Robert F. *The Enemy Within.* New York: Harper & Brothers, 1960.

Los Angeles Times.

Salinger, Pierre, Edwin O. Guthman, Frank Mankiewicz, and John Seigenthaler, eds. *"An Honorable Profession": A Tribute to Robert F. Kennedy.* New York: Doubleday, 1968.

White, Theodore H. *The Making of the President, 1960.* New York: Atheneum, 1961.

PART TWO: MR. ATTORNEY GENERAL

I. The Oral Histories of the John F. Kennedy Library

The Kennedy Library, part of the National Archives, houses the papers of President and Robert Kennedy and of many individuals associated with them. Robert Kennedy's oral history interviews, conducted under the aegis of the Library in 1964, 1965, and 1967, were particularly helpful for this chapter.

II. Books about Robert Kennedy

Guthman, Edwin O. *We Band of Brothers.* New York: Harper & Row, 1971.

Schlesinger, Arthur M., Jr. *Robert Kennedy and His Times.* Boston: Houghton Mifflin, 1978.

III. Other Sources

Kennedy, Robert F. *Just Friends and Brave Enemies.* New York: Harper & Row, 1962.

__________. *Thirteen Days.* New York: W. W. Norton & Co., 1969.

Navasky, Victor S. *Kennedy Justice.* New York: Atheneum, 1971.

PART THREE: THE SENATE YEARS

I. The Oral Histories of the John F. Kennedy Library

Particularly useful for this chapter were the oral history interviews with the following persons:

Cesar Chavez (one volume)
Peter Edelman (eight volumes)
Jeff Greenfield (one volume)
Frank Mankiewicz (nine volumes)

Adam Walinsky (nine volumes)
G. Mennen Williams (one volume)

Generally speaking, unless otherwise indicated, quotations from the above individuals come from these volumes.

II. Oral History Interviews: American Journey and Robert F. Kennedy: In His Own Words

Guthman, Edwin O., and Jeffrey Shulman, eds. *Robert F. Kennedy: In His Own Words.* New York: Bantam Books, 1988. Contains excerpts from the oral histories.

Plimpton, George, ed. *American Journey: The Times of Robert Kennedy.* New York: Harcourt Brace Jovanovich, 1970. Contains excerpts from 347 interviews conducted by Jean Stein with (among others) the friends of Robert Kennedy who traveled aboard his funeral train.

III. Books about Robert Kennedy

Brief quotations attributed to the respective authors have come from the following (these books were also invaluable to our research concerning the contexts of various speeches):

Gardner, Gerald. *Robert Kennedy in New York.* New York: Random House, 1965.

Guthman, Edwin O. *We Band of Brothers.* New York: Harper & Row, 1971.

Schlesinger, Arthur M., Jr. *Robert Kennedy and His Times.* Boston: Houghton Mifflin, 1978.

vanden Heuvel, William, and Milton Gwirtzman. *On His Own: RFK 1964–68.* New York: Doubleday, 1970.

IV. Other Sources

Bedford-Stuyvesant Restoration Corporation: 25 Years of Making a Difference. The silver anniversary retrospective released by the corporation in 1992.

Edelman, Peter. *Searching for America's Heart: RFK and the Renewal of Hope.* New York: Houghton Mifflin, 2001.

Galloway, John, ed. *The Kennedys & Vietnam.* New York: Facts on File, 1971.

Goodwin, Richard N. *Remembering America.* Boston: Little, Brown & Co., 1988.

Kennedy, Robert F. *To Seek a Newer World.* New York: Doubleday, 1967.

__________. *Thirteen Days.* New York: W. W. Norton & Co., 1969.

Kotz, Nick. *Let Them Eat Promises: The Politics of Hunger in America.* Upper Saddle River, NJ: Prentice-Hall, 1969.

Lawford, Patricia Kennedy, ed. *That Shining Hour.*

Hanson, MA: Halliday Lithograph, 1969. A privately printed collection of tributes; attribution is to the individual who wrote the tribute from which a quotation is taken.

Salinger, Pierre, Edwin O. Guthman, Frank Mankiewicz, and John Seigenthaler, eds. *"An Honorable Profession": A Tribute to Robert F. Kennedy.* New York: Doubleday, 1968. Again, attribution is to the individual who authored the tribute.

Scheer, Robert. *How the United States Got Involved in Vietnam.* Santa Barbara, CA: The Fund for the Republic, 1965.

Schlesinger, Arthur M., Jr. *A Thousand Days.* Boston: Houghton Mifflin, 1965.

Shesol, Jeff. *Mutual Contempt.* New York: W. W. Norton & Co., 1997.

Wofford, Harris. *Of Kennedys and Kings: Making Sense of the Sixties.* New York: Farrar, Straus & Giroux, 1980.

Contemporaneous newspaper accounts are generally attributed in the text.

PART FOUR: THE 1968 PRESIDENTIAL CAMPAIGN

Many of the sources indicated for Part Three were also very useful for Part Four, including the Kennedy Library's oral histories of Messrs. Edelman, Greenfield, Mankiewicz, and Walinsky, and the books *American Journey, Robert Kennedy and His Times,* and *Mutual Contempt.* Additional sources for Part Four included:

Aguirre, Michael J., ed. *The Speeches of the 1968 Campaign of Robert F. Kennedy.* Manuscript, 1986.

Halberstam, David. *The Unfinished Odyssey of Robert Kennedy.* New York: Random House, 1968.

Mariscal, Jorge. "Cesar and Martin, March '68." Essay in the Farmworker Movement Documentation Project's collection at the University of California San Diego libraries.

Matthews, Christopher. *Bobby Kennedy: A Raging Spirit.* New York: Simon & Schuster, 2017.

Newfield, Jack. *Robert F. Kennedy: A Memoir.* New York: Berkley Publishing, 1969.

Sann, Paul. *The Angry Decade: The Sixties.* New York: Crown Publishers, 1979.

Thomas, Evan. *Robert Kennedy: His Life.* New York: Simon & Schuster, 2000.

Tye, Larry. *Bobby Kennedy: The Making of a Liberal Icon.* New York: Random House, 2016.

Witcover, Jules. *85 Days: The Last Campaign of Robert Kennedy.* New York: William Morrow & Co., 1969.